SpringerBriefs in Applied Sciences and Technology

SpringerBriefs present concise summaries of cutting-edge research and practical applications across a wide spectrum of fields. Featuring compact volumes of 50 to 125 pages, the series covers a range of content from professional to academic.

Typical publications can be:

- A timely report of state-of-the art methods
- An introduction to or a manual for the application of mathematical or computer techniques
- A bridge between new research results, as published in journal articles
- A snapshot of a hot or emerging topic
- An in-depth case study
- A presentation of core concepts that students must understand in order to make independent contributions

SpringerBriefs are characterized by fast, global electronic dissemination, standard publishing contracts, standardized manuscript preparation and formatting guidelines, and expedited production schedules.

On the one hand, **SpringerBriefs in Applied Sciences and Technology** are devoted to the publication of fundamentals and applications within the different classical engineering disciplines as well as in interdisciplinary fields that recently emerged between these areas. On the other hand, as the boundary separating fundamental research and applied technology is more and more dissolving, this series is particularly open to trans-disciplinary topics between fundamental science and engineering.

Indexed by EI-Compendex, SCOPUS and Springerlink.

Mark Dodgson • David Gann

The Search for Viable Fusion Energy

In Pursuit of the Sun Gods

Springer

Mark Dodgson
University of Queensland
St Lucia, Brisbane, Australia

David Gann
Magdalen College, University of Oxford
Oxford, UK

ISSN 2191-530X ISSN 2191-5318 (electronic)
SpringerBriefs in Applied Sciences and Technology
ISBN 978-3-032-13465-3 ISBN 978-3-032-13466-0 (eBook)
https://doi.org/10.1007/978-3-032-13466-0

This Springer imprint is published by the registered company Springer Nature Switzerland AG
The registered company address is: Gewerbestrasse 11, 6330 Cham, Switzerland

For Max, Noah, Arden, Will, and Ettie

Acknowledgments

A great many people have helped us with this book, as interviewees and commentators. Mark Dodgson would like to thank those people who kindly let him interview them, politely keeping to themselves his breathtaking ignorance. They include Luc Bardin, Tim Bestwick, Damian Brennan, Rob Buckingham, Ian Chapman, Lorne Horton, Lyanne Maclean, Paul Methven, William Morris, and Amanda Quadling. Especial thanks are offered to those who commented on drafts of the book, including Anne Asha, Lindsay Burley, Robin Burley, Norman Harrison, and Frans Berkhout. Luc Bardin's extensive comments and wise advise have significantly improved the book. Heather Lewtas and Peter Fox diligently corrected our science and some of our history. We are grateful for the support of Charles Godfray and the Oxford Martin School.

The authors have presented the research for the book at a number of academic seminars, executive education programs, and industry forums in the UK, Europe, and the USA, and the feedback from these events has been invaluable. Mark would like to acknowledge the generosity of Magdalen College, Oxford, for hosting him during the research for and writing of this book. He would also like to thank Philip Pullman for a lifetime of showing him how stories can hold deeper meaning, how art and science feed each other, and for introducing him to the work of Benjamin Labatut who shows there is no limit to the imaginative telling of science.

David's acknowledgements are deeper and wider due to his roles as previous Chair of the UKAEA and current Chair of UKIFS. He would particularly like to acknowledge his debt to colleagues on the Boards of these organizations and their senior leadership teams, who have deeply influenced his understanding of fusion and how the worlds of science, government policy, and business connect. He has had the opportunity to host numerous visits to the UKAEA's Culham site by Prime Ministers, Ministers, foreign dignatories, business leaders, senior public servants, and eminent scientists, and the range of their rich interests and questions has informed the telling of the fusion story in this book.

In representing the national interest in the roles at UKAEA and UKIFS, David acknowledges that it influences his perspective on fusion, but also recognizes the obligation it demands for objective and impartial judgment on its progress and potential. It also, of course, offers advantageous access to the information included in this book.

He is grateful for the invaluable support provided in his office, including by Lyn Haslett-Hawkins, Alex Cheetham, and Lynn Boswell.

We are grateful for the insights provided by Marina Zhang and Leo Chen on China's approach to fusion energy.

We are very thankful for the generosity of Chris Philpot in giving us permission to use his illustrations of fusion machines.

As ever, with all our research and books, little could be achieved without the unwavering, not to say heroic, support of Sheridan and Anne.

Competing Interests The authors declare the following as potential competing interests: David Gann is Chair of UK Industrial Fusion Solutions and was previously Chair of the UK Atomic Energy Authority

Contents

Part I
Myths and Truths

Chapter 1
The Story to Be Told

I sell here, Sir, what all the world desires to have - POWER. Matthew Boulton, 1776.[1]

We have to evolve means for obtaining energy from stores which are forever inexhaustible, to perfect methods which do not imply consumption and waste of any material whatever. Nikola Tesla, 1897.[2]

This is a book about the pursuit of fusion energy: the attempt on earth to replicate the same physical processes that heat the Sun and the planets to produce viable energy; that is energy that is safe, affordable, and sustainable. The book will explore the complexity and challenge and what we see, for the most part, as the nobility of this quest. The pursuit is happening because of dire ecological and economic need, and the boundless curiosity and inventiveness of humankind. As with the ambitions of all new and alternative sources of energy, it also has to be seen in the light of the pensioner struggling with power bills—choosing between heating and eating—by the rage of climate change protestors, blocking motorways, and defacing works of art, and by those disquieted by reliance on kleptocratic regimes in states rich in oil and gas. The significance of new energy sources also lies with the millions of people in poorer nations whose opportunities in life are curtailed by lack of access to electricity and power. For all its scientific and technological ambitions, in the end the purpose of the pursuit is to provide better lives for people by producing reliable and accessible energy that does not damage the environment or global democracy.

---

[1] Entry for Friday 22 March 1776. Birkbeck-Hill, G. (ed.), *Boswell's Life of Johnson* (1934–50), Vol. 2, 459.

[2] Speech (12 Jan 1897) at a gala inaugurating power service from Niagara Falls to Buffalo, NY. Printed in "Tesla on Electricity", *The Electrical Review* (27 Jan 1897), **30**, No. 3, 47.

M. Dodgson, D. Gann, *The Search for Viable Fusion Energy*, SpringerBriefs in Applied Sciences and Technology, https://doi.org/10.1007/978-3-032-13466-0_1

This book tells of the journey fusion energy[3] has taken, from the unproven theories of some of the greatest physicists of all time, to addressing immensely complex scientific and engineering challenges to prove the feasibility of fusion energy production, to attempt to make the possible practical in commercial energy plants. It is the story of the hard, dedicated work of a wide variety of people, of institutions adapting as they respond to and are buffeted by political headwinds, of incredibly complex and ever-evolving machines, and of government support transitioning from support for science to support for industry. It is a story of the progress of what we shall call fusion innovation ecosystems, and whose actions are intensifying.

The telling of this journey is both prosaic and poetic. It includes the most noble of human characteristics in the aspiration to understand and make the world a better place. It involves the stark necessity to address climate change and the desire of nations to improve energy security and develop industrial capabilities. It has to account for the ambitions of individual scientists and engineers, countries seeking competitive advantages, and firms searching for future profits. So, on the one hand, it is a story of the virtues of inexhaustible human curiosity and creativity, on the other hand it is the story of pragmatic policies and strategies to build fusion machines and a fusion energy industry as part of a global fusion market estimated to range from £3 Trillion to 12 Trillion.[4] Both poetry and prose are needed to describe a fusion machine which is an engineering wonder of complex materials and shapes that creates the awesome power and ethereal beauty of a plasma, superheated at 150 million degrees centigrade, ten times hotter than the centre of the Sun.

A pursuit entails a journey, and in a journey lies a story. The use of stories is much valued as a form of communication in science and business, aiming to evoke the interest and concern of the reader and also, hopefully, helping remember what in the book is important, and why. The story of the pursuit of fusion energy builds on a series of ancillary stories, for example, of the people and institutions behind the pursuit.

The journey involves the attempt to realize all the advantages of fusion energy. These include way its fuels are effectively inexhaustible. Fusion is also highly efficient in the sense that it produces more energy per gram of fuel than any other energy source, while producing no carbon or other greenhouse gases. Unlike nuclear fission, fusion does not involve any chain reactions and can be immediately switched off, and its radioactive waste has a tiny fraction of the half-life of that produced by fission. It also has the benefit, compared to many renewable sources of energy such as solar and wind, that it can continually operate to support baseloads—the minimum required for continuous supply—and can do so close to the point of need.

---

[3] Fusion is a physical process, and hereafter when we refer to fusion energy it is to the controlled use of fusion to produce consumable energy. As its name implies, it involves the fusing together of atoms, a completely different process from fission—the means by which nuclear energy is currently harnessed—which involves splitting atoms. This profound difference, as we shall see, has significant implications, making, for example, the process of fusion energy production much safer (see Chap. 4 especially).

[4] Written evidence submitted by UKAEA and UKIFS (IND0009) to UK Parliamentary Committee on Science and Technology, February, 2025. Chromeextension://efaidnbmnnnibpcajpcglclefindmkaj/https://committees.parliament.uk/writtenevidence/137642/pdf.

The challenge is to make the highly complex story about this journey easy to navigate. Fusion energy is a scientific experiment; a technological, engineering, industrial, and commercial project; and a political and policy initiative. Its ambition has cultural significance, and it is a testament to humankind's endless creativity and determination. It is all these at a scale seldom matched in history. To tell this story, we take the advice of the master story-teller, Philip Pullman (2017), when he says the writer's attention should be with the path and not the woods. While the woods are full of fascination—and certainly the story of fusion is full of related and consequential influences, in its science, engineering, and economics, for example—our aim is not to be distracted and to resist them: to be informed by detail, but not swamped by it.

Stretching the analogy, the writer also needs to see the wood for the trees, and this involves personal choices on what is important and what is not: choices that others may not make. This leads us to write about the motivations for going down a path, and the reasons why the path has taken the course it does, which in turn needs to know where it started: hence, the necessity of some history. Sustenance is needed to pursue the path, as is lighting when it gets dark and uncertain, and these require significant resources, and this poses the questions of whether these are sufficient for the purpose and what the cost will be if they are insufficient. A few believe the path will be easy, more say it will be rocky but passable, and some believe it leads to a dead end. Furthermore, there may be more than one path to reach the destination, and its direction may be altered with the discovery of new information or tools for searching its route. With all these questions and views in mind, and apologies for flogging the analogy to death, our aim is to focus on the quest, why it is important and hard, the struggles it faces, and why a great deal hangs on the outcome of the story: a story that is continuing and is yet to reach a conclusion. The destination of the journey is the provision of safe, secure, and competitively priced energy, but as we shall see there remain many obstacles to be overcome before that destination can actually be determined and reached.

Amongst the greatest stories, of course, are the ancient myths, and our story connects with many of them in their focus on the Sun and the blessings it brings. Helios, the Greek god of the Sun, is a deity in the pantheon of classical mythology known for the golden chariot he drove each day pulling the Sun from east to west, bringing Sunshine and life to earth. Helios was especially worshipped in Rhodes and he was depicted by the Colossus, the statue which stood at the entrance of the harbour of the Greek city in the early third century BC. Approximately 33 m high, the same height as the Statue of Liberty, and in existence for just over half a century before it was destroyed by an earthquake, the magnificent statue in iron and bronze is regarded as one of the Seven Wonders of the Ancient World.

Helios's chariot that moved the Sun was pulled by four horses, including Pyrois (the fiery one), Aethon (blazing), Phlegon (burning), and Aeos (he of the dawn). The journey was occasionally disrupted, most notoriously when Helios's son commandeered the chariot and lost control of it, causing parts of the Earth to catch fire and invoking the fury of Zeus, King of the Gods. Other gods, their births and deaths, sporadically caused delays, and Helios sometimes lingered in the morning with his wife, causing shorter days, and became distracted on his journey by nymphs, creating longer days.

Helios and his gift to the world of the Sun, the immensely hot ball of plasma in the centre of our solar system, provides a metaphor for the attempt on Earth to replicate its source of energy, bringing the Sun to Earth in the form of nuclear fusion. To achieve this, and the huge temperatures required, is also a vital journey fraught with challenges. The chariot moving the Sun across the sky is pulled by fierce and powerful horses needing extraordinarily skilful handling. Led badly, it crashes and burns. Its journey can be curtailed by unpredictable forces and distracted by the easy comforts of the familiar and tempting but elusive diversions. This book resonates with this metaphor and is the story of a modern pursuit representing the nature and form of the chariot, and how the blazing, burning, and fiery forces that propel it are created and directed (and perhaps creating a new dawn for energy production). The ambitions and edifices surrounding the making of fusion energy mirror those of the Colossus of Rhodes in its era as the effort strives to become a Wonder of the Modern World.

Myths are metaphors that usefully help compare different things based on similarities. Greek gods brought the Sun to Earth; fusion energy aims to replicate the physics of the Sun and stars on Earth. While the scale of ambition is different: without daily sunshine there would be no life on Earth, the aim of producing clean, sustainable energy through fusion energy aims to contribute to the maintenance of life on Earth. The forces carrying the Sun gods across the sky are powerful and challenging to control, and the delivery of fusion energy on Earth involves overcoming some of the most difficult scientific and engineering problems of all time. Just as the gods could become distracted, so the journey to produce fusion energy can be disturbed and redirected by promising but ultimately unviable alternatives, and hubris shadows all great endeavours: the lessons of Helios's feckless son are always to be remembered. Similarly, the myth of Icarus, the wax of whose wings melted as he flew too close to the Sun, is often used as a morality tale: beware of over-exuberance and unfettered ambition.

Metaphors can only be stretched so far, and connecting ancient Greek myth with modern science and technology may seem to be pushing the boundaries. But myth and tradition play a crucial reference point for the present. Myths continue to have an impact on today's society, and as Helen Morales (2007:4) argues, they matter "for us in understanding who we are: our selves, … and our lives".

Just as Philip Pullman argues science offers useful metaphors for the arts, the philosopher of science Karl Popper (1957) argued that science must begin with myths, and with the criticism of myths. He believed that myths are important sources of knowledge and science originates from traditions, but that we should question them, accept the ones we find useful and reject those we don't understand. For our purposes, the connection is simple: how the pursuit of fusion energy resonates in some ways with the epic journeys of the Sun gods. We might also recall George Orwell's view that myths that are believed in tend to come true.[5]

[5]Davison, P., Angus, I. and Davison, S. (eds) (1998), *The Complete Works of George Orwell: I have tried to tell the truth, 1943–1944*, Secker and Warburg, Indiana University: p24. A startling example is provided by Marathon Fusion, a US start-up that claims by using a fusion process of

And the journey towards viable fusion energy—energy that is high performing, consistent, reliable, and cheap—is truly epic. Scientists have searched for and studied fusion for 100 years, technologists and engineers have strived to build a machine to produce it, and politicians and businesspeople have wished they could make it happen. If it were easy, they would have succeeded and we would have one means of helping ameliorate the climate crisis that threatens our very existence. The determination to produce consumable energy from fusion has induced international collaboration between nations more attuned to conflict. Governments and private businesses, convinced of the potential of fusion energy, are investing billions of dollars into making it a reality. Yet, in stark terms, after decades of research and billions of dollars invested, fusion has yet to report a net gain of more energy out of a machine than the total amount that is put in it.

The importance of the outcome of the pursuit is clear. The earth's climate is being damaged so severely by the type of energy we use that our biosphere is being destroyed and we may not be able to survive in many areas that were previously habitable. There is scepticism about the ability of current plans to decarbonize and reach net zero greenhouse gas emissions by 2050, with devastating consequences for the world. The world urgently needs all possible sources of safe, clean, and carbon-free sources of energy. It is hoped, but by no means guaranteed, that fusion can contribute as such a source of energy.

This book contends that, as well as this crucially important objective, the search for fusion energy provides an example of humanity at its best, a journey of optimism and deep expertise. It tells of the potency of basic research and its connection to advanced practices in engineering, construction, and computer and data science. It puts a spotlight on the importance of experiments, some of which have to be on an enormous scale. It shows how knowledge develops iteratively, sometimes quickly, but more often with many setbacks and failures. It is a story of wonder, perseverance, and intellectual effort and ambition, directed at solving a problem of global significance. Its narrative is an antidote to scientific scepticism, parochialism, and insularity; its origins and progress are typified by internationalism and institutions with long-term orientations preparing for a world decades ahead.

And it is a story of connections and collaborations. In many ways, the pursuit of fusion energy, the fusion of atoms, is an act of fusion between very different parties, including the institutions of science, departments of government, and organizations in business and finance. It has involved the fusion in an innovation ecosystem of the ambitions of research establishments, public organizations and policies developed to improve energy sources, and the investments and activities of for-profit firms.[6] At the personal level, the pursuit of fusion energy has progressed from the minds of the founders of modern physics—Curie, Einstein, Rutherford—to modern leaders of science, working at the forefront of understanding issues such as the control of plasmas and new materials that can resist extreme temperatures, onwards into the challenges faced by engineers producing machines of awesome complexity, and latterly

nuclear transmutation it can produce gold, potentially delivering the millenia-old myth of alchemy. (https://www.ft.com/content/06f91e0d-3007-40bd-b785-86fef4890809).

[6] There is a growing academic literature on innovation ecosystems. See, for example, Granstrand and Holgersson 2020.

engaging with managers in companies and investors adjudging risks in order to make profits. It is essentially an act of fusion of very different sciences and technologies: from the basic science of physics to advanced engineering of materials and robotics, to the use of cutting-edge digital technologies to design fusion machines and their control systems. Each of these fields is further pushed to their extremes through their need to connect with one another in the design and manufacture of massively complex machines and their connections to power systems.

The central role of governments has changed in this pursuit: from investing in basic science, to incentivizing and re-risking public and private investments in fusion machines with the aim of producing the social and economic benefits of fusion energy. The ways governments are pursuing fusion reflects their political ideologies: China is state-led, with increasing private sector engagement; the USA looks to private sector firms connected back into the science base; the UK plans to leverage its scientific strengths by building supportive industries. Beyond this, the story is so rich it needs to be told in a manner that captures its wider essence of the connection between the quest to understand and solve problems with the existential desire to survive and thrive: to create sources of energy that do not destroy the planet.

The pursuit has not been smooth; human frailties and delusions, inflated expectations, disorganization, and political differences and arguments continually intervene. They could but not: the interaction of people, organizations, politics, and money around ambitious and risky innovations will always cause stresses, and, given its significance and potential, these are accentuated in the case of fusion. This is also a story of human egos and frailty, mis-organization and mismanagement, government ineptitude and indecision. It is a tale of chance events, most vividly seen in a murderous airline hijacking, and unlikely situations such as when UK scientists helped the USSR's nuclear research programme during the height of the Cold War. The advances of rational and dispassionate science and engineering are inevitably both stimulated and bounded by the awesome capacities and continuing limitations of the human condition. The fusion of the worlds of science, and its quest for knowledge, governments, and their need to satisfy citizens, and business, and its quest for profit, is not easily achieved.

There is one other association with the idea of myth. For many years, it was held that fusion energy was itself a myth, with the saying that its delivery was decades away and always will be. The production of fusion energy is no longer a myth, and any mysteries that remain are less to do with its scientific and technical feasibility, although major challenges still remain, and more to do with its timescales and economics. We emphasize, however, that we are writing about a continuing journey rather than arrival at a destination: fusion has yet to power any commercial machine or feed an electricity grid. Our story tells of a path whose direction is becoming increasingly clear, but whose journey's end remains to be determined. We shall see some claims, mainly by small fusion firms, that fusion energy is just around the corner. More realistically, the sheer complexity of the science and engineering challenges and the constant and inevitable political and organizational trials surrounding fusion energy lead some of the most informed fusion advocates to predict that fusion is very unlikely to make a significant contribution to reducing CO_2 production much before 2050 (Chapman et al. 2024). As we shall see, there are many related benefits that have been and will be achieved much more quickly from investments in fusion energy, and whose timescale for realization will also depend on the level of resources dedicated to it.

The book's core is comprised of two major elements: the stories of people involved, their contributions and motivations, and the story of the innovation journey in the pursuit of fusion energy. Analysing the journey through the lens of innovation helps provide some intellectual scaffolding for the book. Innovation is defined as the successful application of new ideas that add value, and it is both an outcome and a process. As an outcome we focus on the creation of fusion machines, and we shall explore the innovation process as it proceeds from science, "new ideas", to technology, "successful application", and commerce: "add value". As a process we highlight how progression in the development of fusion energy has involved a great deal of learning and collaboration: key features of innovation. We shall argue that the further coalition and effectiveness of fusion energy's innovation ecosystem is key to realizing fusion's potential. As an indicator of the future direction of the journey of fusion energy, we shall use the example of the development of a fusion energy prototype plant as a catalyst designed to accelerate the cohesion and improve the performance of an innovation ecosystem. First, we briefly explain fusion energy.

References

I. Chapman, S. Cowley, H. Wilson, The spherical tokamak for energy production: theme issue introduction. *Phil. Trans. R. Soc.* **A 382**, 20230416 (2024)

O. Granstrand, M. Holgersson, Innovation ecosystems: A conceptual review and a new definition. Technovation **90**, 102098 (2020)

H. Morales, *Classical Mythology: A Very Short Introduction* (Oxford University Press, Oxford, 2007)

K. Popper, The philosophy of science, in *British Philosophy in the Mid-Century*, ed. by C. Mace, (George Allen & Unwin, London, 1957), pp. 155–194

P. Pullman, *Daemon Voices: On Stories and Storytelling* (David Fickling Books, Oxford, 2017)

Chapter 2
A Short Introduction to Fusion and Its Machines

British Prime Minister Winston Churchill had the highest regard for scientists. Science was crucial to the winning of the Second World War, seen, for example, in the development of radar, code-breaking computers, and the ultimate weapon of the atomic bomb. Churchill's enthusiasm and support for science, which had such an influence on the outcome of the War, predated the global conflict. He credited science as the cause of what he described as the prodigious speed of human advance. Following its early scientific discoveries, the potential of fusion energy was quickly realized by Churchill, who wrote in 1931 that:

> If the hydrogen atoms in a pound of water could be prevailed upon to combine together and form helium, they would suffice to drive a thousand horsepower engine for a whole year... If those electrons, those tiny planets of the atomic system, were induced to combine with the nuclei in the hydrogen the horsepower liberated would be 120 times greater still. There is no question among scientists that this gigantic source of energy exists.[1]

Although he was alert to its dangers, he was enamoured of the potential of fusion, saying that:

> The discovery and control of such sources of energy would cause changes in human affairs incomparably greater than those produced by the steam-engine four generations ago. Ibid

Approaching a century after Churchill's observations, the potential of fusion energy has been demonstrated, but it has not advanced at prodigious speed and it has yet to cause any changes in human affairs by producing viable energy in new ways. There are questions to be asked of whether Churchill's vision was right all those years ago. Our suspicion is that if he were to view fusion energy from the vantage of today, his enthusiasm would be undimmed. Given his respect for scientists, he may share the view of Professor Stephen Hawking who, when asked to describe the one idea that would transform our society, answered: "I would like to see the

[1] Winston Churchill, *Fifty Years Hence,* December 1931. www.nationalchurchillmuseum.org.

M. Dodgson, D. Gann, *The Search for Viable Fusion Energy*, SpringerBriefs in Applied Sciences and Technology, https://doi.org/10.1007/978-3-032-13466-0_2

development of fusion energy to give an unlimited supply of clean energy... Nuclear fusion would become a radical energy source and would provide us with an inexhaustible supply of energy without pollution or global warming".[2] Or, more boldly, the view of Professor Brian Cox that "fusion is the future of our civilization".[3]

Turning such enthusiastic vision into reality depends, however, on the practical creation of machines that generate and harness fusion energy safely, sustainably, and economically. The adage attributed to Thomas Edison that vision without execution is hallucination, may seem unduly harsh and limiting, but Churchill's championing of science was essentially based on its successful practical application.

2.1 Some Basic Principles of Fusion Energy

The evidence of fusion is all around us; without it there would be no Sun and no us. The basic principles of fusion energy can be summarized as follows: Atomic nuclei are positively charged and repel each other, but they can fuse if they collide fast enough to overcome the force that repels them. Very high temperatures are needed to gain that speed, commonly in fusion experiments in excess of 150 million degrees Celsius. At such high temperatures, nuclei and electrons are separated and create plasmas of free moving negatively charged electrons and positively charged nuclei that combine together. The plasma is extremely unstable and difficult to control although sophisticated computer simulation and modelling now helps predict some of its behaviour. It is also so hot that it cannot contact any surfaces and in one kind of fusion machine, the tokamak—which uses a toroidal or doughnut-shaped vessel and is the most common means of creating fusion around the world—the heat is contained in chambers by powerful magnetic fields that confine the plasma. Electric currents in magnetic coils hold the hot plasma away from the walls of the vessel. The most common fuels used in tokamaks are isotopes of hydrogen, tritium, and deuterium, which when fused create a helium nucleus and a neutron. The neutron is harvested for energy production.

Deuterium is found in seawater; tritium is very rare but can be manufactured from lithium, which is also abundant, although its manufacture, as we shall see, is very challenging and complicated. The reference is made that fusion energy utilizing the raw materials found in one bath of water and two computer laptop lithium batteries will produce as much energy as 40 tonnes of coal and 250 kilograms of uranium ore, and it could produce one person's energy use (assume 20 kWh/day) for 60 years. According to one estimate, there is sufficient deuterium to supply the Earth's current energy demand for 100 billion years and sufficient lithium from which to produce tritium for 60 million years (Irvine 2011). However, this

[2] Reported by the BBC, 17th November 2016. https://www.bbc.co.uk/future/article/20161117-stephen-hawking-why-we-should-embrace-fusion-power.

[3] Brian Cox. Horizon—Can we make a star on earth? 28th February 2009, https://www.bbc.co.uk/programmes/b00hr6bk.

significantly underplays the challenges of isolating these materials. Tritium is particularly difficult to make and currently requires production in nuclear fission plants: approaches to producing tritium are discussed in Chap. 6, including efforts to create a virtuous cycle of it being harvested as it is used.

Marie Curie said she was taught that the way of progress was neither swift nor easy,[4] and some of the extreme scientific challenges of harnessing fusion energy are described in the *Bulletin of the Atomic Scientists* (Drollette 2024):

> In a nutshell, the fusion process releases neutrons that are 10 times more energetic than what a commercial plant powered by the splitting of atoms, or nuclear fission, ordinarily emits. These high-powered neutrons are difficult to contain and rapidly degrade the containers proposed for controlling the extremely hot plasma required for a fusion reaction. At the same time, plasmas are just plain difficult to keep stable while producing that all-important steady (or quasi-steady) fusion "burn".

The broader challenges are described by Frans Berkhout, Professor of Environment, Society and Climate at Kings College London:

> The fundamental problem of the huge power needed to achieve initiation and the remaining materials, systems and operational issues in operating for longer periods is clearly the nub of the linked scientific, engineering and management problems.[5]

In addition to the challenges of heating, containing, and controlling plasmas in tokamaks is the profound engineering problem of how to transfer the extreme heat being produced in the plasma to generate energy. Only around 40 percent of the heat produced can be harvested as energy. The neutrons being produced in the fusion reaction deposit their energy on the walls of the vessel and need to be used before they melt them. The magnetic field that contains the plasma in tokamaks is produced by coils that are cooled by liquid helium or liquid nitrogen cryogenics. A challenge then is to use materials for the walls that will absorb the neutron energy and allow the heat to be conducted away to generate electricity by, for example, using conventional turbine machinery while protecting the cooled magnets. In this way, rather remarkably, some of the most advanced technology of this century connects with the technology of steam engines developed in the eighteenth century.

2.2 Fusion Machines

The awesome capacities of machines have always excited and intrigued. Leonardo Da Vinci's genius included his designs of helicopters, robots, and self-propelled carts in the fifteenth century. Written in the nineteenth century, Jules Verne's submarine *Nautilus* and his cannon *Columbiad* created to fire astronauts to the moon are joined with HG Wells' Time Machine as machines that captured the popular imagination. Not always welcome, remembering the Luddites breaking textile machines

[4] https://history.aip.org/exhibits/curie/brief/06_quotes/quotes_03.html.

[5] Berkhout, F., personal communication, 26th January 2024.

that threatened their livelihoods, they have fascinated even the most otherworldly. Upon seeing their first industrial machine, the Romantic poet William Wordsworth thought it to be like a creature "making the first step from brute matter into life and purpose, showing its progress by great power", and his friend and fellow poet Samuel Taylor Coleridge observed that "it seemed like a giant with one idea". (Quoted in Uglow 2015). Standing beside a tokamak today—a machine of extraordinary size and complexity, and capable of producing such extremes of temperature—induces the same sense of awe.[6]

The etymology of the word fusion, from Latin and Ancient French, lies in blending and uniting different things by melting them through heat, and therein lies the purpose of a fusion machine: the merging of atoms at extreme temperatures. The challenge is to create, channel, control, and scale fusion reactions repeatably, sustainably, and economically. A wide variety of machines have been made to do this. Some have enjoyed exotic names, such as the Stellarator, Perhapsatron, Pyrotron Mirror, and Z-Pinch. Others are more prosaic: tokamak is a Russian acronym of toroidal chamber with magnetic coils. While different kinds of fusion reactors aim for the same very high temperatures, they adopt a variety of approaches to do so.

The idea of creating an actual tokamak emerged from the work of two Soviet scientists, Igor Tamm and Andre Sakharov, in 1950–51, although in his memoirs Sakharov says he was thinking about the problem of controlled thermonuclear reaction from 1949 (Sakharov 1990). He also relates how his thinking was stimulated by a young soldier, Oleg Lavrentiev, based with the USSR's Pacific Fleet, who had proposed an important source of energy production through the use of an electrostatically confined high-temperature deuterium plasma.[7] Tamm and Sakharov knew that electrostatic control was unfeasible, and suggested instead that the confinement of the plasma should be by means of magnets. They researched the shape of the toroid, behaviour and control of unstable plasmas, use of deuterium and tritium as fuel sources, and the problem of heat dissipation. Andre Sakharov is best remembered today for being an eminent dissident and winner of the Nobel Peace Prize in 1975 for his promotion of human rights, but it is also to be recalled that the design of every present and future tokamak in the world can be traced back to his work in Moscow on the cusp of the 1940s and 1950s. Indeed, the idea of a tokamak can be tied back to an entirely self-taught 24-year-old soldier working as a signalman in the wilds of Sakhalin Island in Pacific USSR.

There are now over 60 different tokamaks in operation around the world. The tokamak design has evolved as experiments create more reliable knowledge about what might work cost-effectively. Fusion energy's penetration of the electricity markets will ultimately be determined by economics—whether machines can produce sufficient, reliable, safe electricity to compete on price with alternative sources.

[6] ITER's tokamak, for example, will be 29 metres high and 28 metres in diameter, weighing 23,000 tonnes. It is literally a Colossus.

[7] Sakharov was to meet Lavrentiev in person in the offices of Lavrentiy Beria, the terrifying Chief of Stalin's secret police, where Beria quizzed them about fusion energy. Lavrentiev did not fit in with Moscow's physics community and spent his career in a small physics department in Ukraine.

The major costs of fusion power plants are the capital costs of building the facilities and the magnets and cryogenics equipment, and the cost of finance. Reducing these costs will be key to fusion's commercial viability, and one way to achieve this is to reduce the whole scale of the plant and equipment, and as we shall see efforts are being made to reduce the size of tokamaks including in more spherical shapes ("apple" rather than "doughnut" shaped) although these will themselves be of significant size. Figure 2.1 illustrates the main components of a tokamak.

The many challenges to be confronted include the huge amount of power needed to create the burning plasma and then ensuring its stability and continuity at very high temperature. The neutrons produced in the fusion reaction are highly damaging to materials they come into contact with, so creating resilient materials in any containing vessel is crucial. Fuel self-sufficiency is a major challenge. As vessels have to withstand enormous heat and will be radioactive, maintenance is necessary and has to be conducted remotely, and robotic maintenance will need to be highly sophisticated. The enormous heat exhaust and helium "ash" generated has to be

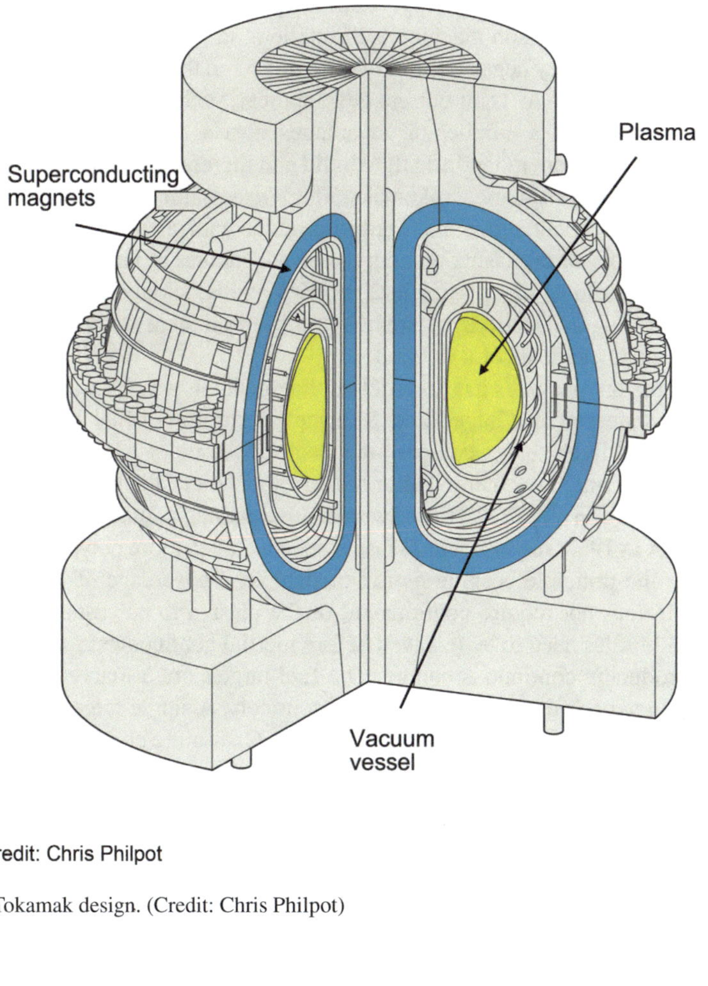

Credit: Chris Philpot

Fig. 2.1 Tokamak design. (Credit: Chris Philpot)

utilized and managed. As we shall see, in contrast to existing fission nuclear plants, this is eminently doable. Superconducting magnets producing very low temperatures are used to contain the heat, but these can be exposed to what is known as the quench phenomenon, which refers to the sudden loss of superconductivity in a material, often leading to a rapid increase in resistance and a rise in temperature. This is a critical event in superconducting magnet systems, where maintaining low temperatures is essential for their operation. Mastering the physics of such behaviours is one of the main remaining technological challenges in creating fusion machines. The complexity of each fusion energy facility also requires use of data analytics, and digital engineering design, integrated with advanced engineering and manufacturing. And all this needs to be done in designs that are manufacturable at costs that deliver energy at viable prices.

As of 2025, the record for maintaining a steady-state, high-confinement plasma, is held by China's Experimental Advanced Superconducting Tokamak (EAST) the Institute of Plasma Physics at the Institute of Physical Science of the Chinese Academy of Sciences on Hefei's "Science Island". It maintained a plasma for over 17 min and is seen as a significant breakthrough.[8]

Another kind of fusion machine is the stellerator, which are toroids twisted into a figure of eight. They began to be explored in 1951 in the Princeton Plasma Physics Laboratory in the USA. Their current development, such as Germany's Wendelstein 7-X Stellarator, has a number of advantages over a tokamak but their complex geometry makes them more difficult to build and therefore they have not progressed as fast as toroidal or spherical tokamaks. The Wendelstein 7-X is the world's largest stellerator experiment and is expected to produce equivalent performance to tokamaks in the control of plasma without the disadvantages of large electricity current flows. It produced its first plasma in 2015. Essentially, tokamaks are better at creating hot plasmas, and stellerators are better at keeping them stable. Figure 2.2 illustrates the main elements of a stellarator.

Other fusion machines use "inertial confinement" in which fusion fuel is instantaneously squeezed by a huge force. Miniscule quantities of fuel are bombarded by high-powered laser beams compressing them to the point where fusion reactions occur. This technology emerged after the development of the laser in 1960 and was first used in fusion research at the Lawrence Livermore National Laboratory (LLNL) in the USA in 1962. It was not until lasers became much more powerful two decades later that the principle became more practical. The advantage of inertial confinement is it does not require confinement of the plasma to be sustained. But major technical hurdles need to be resolved before inertial confinement reactors are capable of producing continuous energy. The fuel targets are a marvel of engineering design and very difficult to manufacture accurately. A single target can take weeks to design and make into an experimental target. One future challenge for producing useable energy from an inertial fusion machine is the requirement to produce something like 800,000 fuel targets per day. A second challenge is to accurately and

[8] https://phys.org/news/2025-01-chinese-artificial-Sun-fusion-power.html.

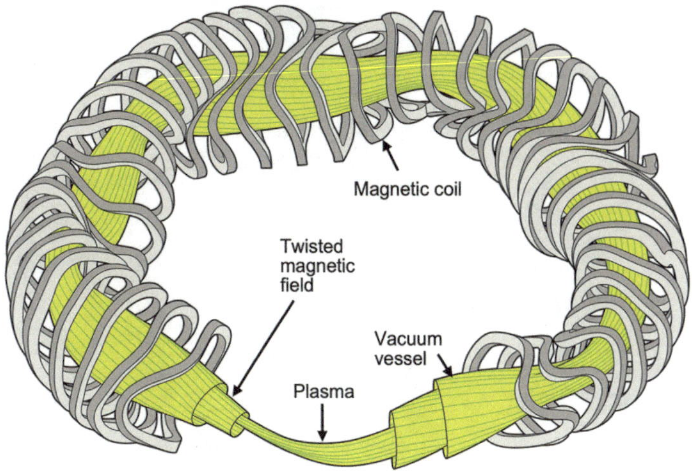

Fig. 2.2 Stellarator. (Credit: Chris Philpot)

continuously fire these fuel capsules in rapid pulses into the target area of the machine, aligned to be hit instantaneously by the enormously powerful lasers, in order to produce an almost continuous process of fusion. A third challenge, common to all types of fusion machine, is the production of tritium fuel, with a potential tritium breeding system combined as part of the machine. Figure 2.3 illustrates the main element of an inertial confinement experimental machine.

There are major inertial confinement research laboratories at the US National Ignition Facility (NIF)[9] at the LLNL in California and at the French Laser Megajoule Laboratory near Bordeaux. The NIF uses the world's most powerful lasers to compress the fusion fuel. It is housed in a ten-storey building and the lasers are amplified in circuits across an area the size of three football pitches, finally hitting the target with 192 beams, delivering nearly two million joules of ultraviolet laser energy in billionth-of-a second pulses. In feats of extraordinary science, Turrell (2021) describes one such nano-second long experiment as involving: the collapse of a capsule to a thirtieth of its original size at a pace in excess of 350 km a second, at a pressure 300 billion times what we experience on Earth, and involving ten million billion reactions. (4) On the 5th December 2022, the NIF demonstrated fusion ignition for the first time, where more energy from fusion was produced than the amount of energy used to hit the target. This experiment produced 1.5 times more energy than the amount used to create the reaction. This is known as "ignition": although far less energy was produced than the total amount of energy drawn from the electricity grid to produce the lasers and conduct the experiment. The total yield was

[9] https://lasers.llnl.gov.

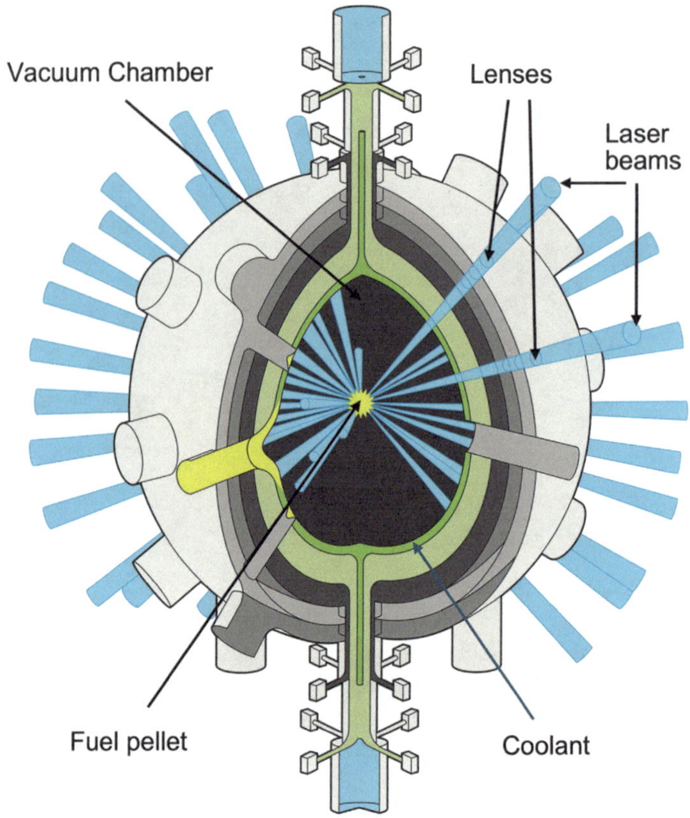

Fig. 2.3 Inertial confinement design. (Credit: Chris Philpot)

over 3 megajoules. By April 2025, subsequent experiments had created a yield of 8.6 megajoules. Modelling predicts that an energy gain of 150–200 megajoules or more will be required to produce a viable inertial fusion energy plant (Plfeugers-Peters 2025). LLNL recognize that harnessing Inertial Fusion Energy requires a massive undertaking and, supported by the US Department of Energy, it has established the Livermore Institute for Fusion Technology, in pursuit of this goal.

All the challenges confronting fusion machines are magnified because if fusion energy is to be produced in viable machines they need to be addressed concurrently. As Chapman, Cowley and Wilson (2024) put it:

> Fusion is different from most other technologies in that the validation of components in a representative environment is only possible in a complete fusion device and probably a complete plant. Furthermore, the cost and timescale of each step means that working towards the final commercial product via a succession of small-increment full physical prototypes is unacceptable, given the urgency of addressing climate change. (Chapman et al. 2024:2)

And all these complex machines comprised different components and technologies at varying levels of development face the difficulties identified by Thomas Hughes (1983) as "reverse salients". These are elements of a system—in his case electric power—that held back progress. When development is uneven, he argues:

> …inventors, engineers, and other professionals dedicate their creative and constructive powers to correcting reverse salient so that the system can function optimally and fulfil system goals. (Hughes 1983:14).

Essentially, the massive complexity of the technologies combined into fusion machines is compounded by the challenge of their integration and uneven state of development.

We now briefly describe three tokamaks: the world's most successful research tokamak (JET), and its decommissioning after 40 years of operation; the world's largest and most ambitious research tokamak (ITER), currently being built and facing significant challenges; and a prototype commercial tokamak (STEP), presently being designed for operation by 2040.

2.2.1 The Joint European Torus (JET)

The world's leading tokamak to date, the Joint European Torus (JET), is located at the UK Atomic Energy Authority's (UKAEA) Culham Campus in Oxfordshire. The village of Culham, with a population of a few hundred people, is known to have existed since the ninth century and one of its claims to fame is a battle in the English Civil War—the battle of Culham Bridge in 1645—at which the Parliamentarian forces defeated the Royalists. In another, because of JET, it has periodically been the hottest place in the Solar System.

Some of the history of JET and its institutional home, and its contribution to a fusion innovation ecosystem, are described in later chapters, and here our concern is with the machine itself. The construction of JET began in 1978 with a life expectancy of roughly a decade. Aided by a series of upgrades it operated from 1983 to 2024. Plans around 2006 to close it down—in the belief that given the development of ITER it would be unnecessary—were curtailed when the extent of the delays confronting ITER became clearer. The tokamak was redesigned so significantly during it lifetime that to all intents and purposes it was a different machine at the end of its life than at its beginning. The continual updating of the technology of subassemblies and subcomponents led to accumulated knowledge on the performance of the whole system. Figure 2.4 is a photograph of the inside of JET with the image of a plasma superimposed.

Testament to its original design, the complexity of the machine and its operation, involving tens of thousands of different systems, evolved over the years to conduct experiments that have been instrumental in establishing the value of the tokamak design and in demonstrating the efficacy of the reactions from the combination of

Credit UKAEA, Eurofusion, https://euro-fusion.org/

Fig. 2.4 JET interior with superimposed plasma. (Credit UKAEA, Eurofusion, https://euro-fusion.org)

deuterium and tritium. It led to a number of significant breakthroughs for the management of plasmas in the vessel and the dispersal of its excess heat.

JET staged a preliminary tritium experiment in 1991, and its first deuterium-tritium experiments using larger amounts of fuel took place in 1997, and it subsequently produced a number of world records. In its final experiment in 2024, 69 megajoules of fusion energy was consistently produced for 5 seconds, using 0.2 mg of fuel. It gained 65 percent of inputted power: 16 MW from 25 MW. JET's world record at the time—which doubled the previous record—still sees less than two-thirds of the power put into the machine produced from fusion, and 69 megajoules is enough energy to boil 70 kettles.

JET comprised eight segments, like an orange. A ninth of these very complex engineered segments was placed in the work hall adjacent to JET where mock up segments were built around it to create a full-scale model of the JET vessel. This assisted the research, development, and operations of remote access handling technology and provided a site for training people to use remote access.

Around 600 people worked on operating JET, and it was such a large and complex machine that it is reputed to take about 8 years to develop the embedded knowledge needed to operate it. The control room during experiments of pulses of plasma created in the vessel involved people and teams controlling power and liaising with the national grid, heating drivers, managing the plasma, and overseeing and recording sensors, all under the oversight of the person controlling that day's experiment. Over its 40-year history, JET delivered 105,929 pulses.

As JET used copper coil magnets, rather than the high-temperature superconducting magnets available today, it used a huge amount of energy to turn them on. Each pulse could involve between 500 and 1000 megawatts, equivalent to the hourly energy requirements for up to two million homes, or 2 percent of the UK's energy supply. Drawing power from the National Grid has a limit of 575 megawatts, so the power surges needed to ignite the plasmas exceeded the Grid's capacity to deliver it. To deal with this, JET used two massive 775 tonne flywheel generators, spinning at 225 revolutions per minute, to store the energy needed for the short periods of the plasma pulses and provide approximately half of the energy needed for a pulse.

Based on the knowledge derived from JET, since the 1990s the UK has pioneered developments of spherical tokamaks, apple-shaped, with a slim core. These are typically smaller than toroidal machines, and are aiming at becoming cheaper, and more reliable. One advantage of spherical tokamaks is that they can use smaller magnets, making more efficient use of the magnetic fields, thus reducing the size and cost of machines. However, the smaller size also presents a challenge in managing the heat in a smaller space. Significant advances have been made in dealing with the redundant heat in smaller tokamaks with the design at the UKAEA of a machine called the Mega-Amp Spherical Tokamak (MAST) which proved the initial scientific theory for the smaller apple-shaped machine. This was upgraded to test a novel design of heat management system, MAST-U, with the results proving the predicted modelling correct. MAST-U involved the design and production of a high complex and technically challenging exhaust system, called the Super-X divertor. It reduced the heat of the exhaust by a factor of ten. This award-winning design could result in a breakthrough in the ability to sustain the very high-temperature plasmas required in a tokamak to provide continuous fusion energy.[10]

JET is in many ways the precursor and progenitor of ITER whose design has continually been influenced by research at Culham, and it is informing the design of STEP; two fusion machines described below. Amongst other lessons, JET has informed their development about the design of the lining of the internal walls, robotic maintenance, and management of spent tiles on the walls, diverting heat from spent plasma, use of sensors to measure the behaviour and heat of plasmas, and methods for heating the plasma. In 2024, JET began the process of being decommissioned, which is leading to valuable lessons for the whole-of-life design of tokamaks. The wide range of UKAEA activities, and its contribution to a fusion innovation ecosystem, is discussed in Chap. 6.

[10] https://raeng.org.uk/news/major-project-award-for-team-behind-machine-vital-to-developing-commercial-power-from-nuclear-fusion.

2.2.2 International Thermonuclear Experimental Reactor (ITER)

ITER is one of the most complex scientific, engineering, and construction projects ever attempted. ITER is a fortunate acronym as *Iter* is also Latin for pathway or the way (to fusion power), nicely emphasizing how viable fusion energy is a pursuit or journey.

Located in Cadarache in the South of France, ITER aims to inject 50 megawatts of thermal power into a tokamak to produce 500 megawatts of thermal output power. It is being built on a site the equivalent of 50 soccer pitches, involves over one million components with ten million parts, and requires from 110 MW to 620 MW of power, with the latter enough to power 1.25 million homes for an hour.[11] Its central tokamak reactor is planned to be ten times larger than any other fusion reactor and is one of the most complicated machines ever designed. It will weigh the equivalent of three Eiffel Towers and will have 100,000 km of superconducting coils (with strands a kilometre long), creating a magnetic field strength three million times the strength of the Earth's magnetic field. The magnet's designers have calculated that their forces can reach 60 meganewtons or twice the thrust that a NASA Space Shuttle requires for lift-off. The tokamak will be contained in a 30x30 metre cryostat, and within a space of 3 metres there will be temperatures of 150 million degrees centigrade and − 269 degrees centigrade. Design of the facility on the CATIA computer-aided-design system involves two trillion bytes of data, and its construction requires two cranes that can lift the equivalent of two fully laden Boeing 747 s. The essential aim of ITER is to demonstrate that a stable plasma can be created to ignite a fusion reaction. Figure 2.5 is an aerial photograph of the especially constructed 42 acre platform for the ITER tokamak and associated buildings.

The history of ITER, the significant challenges it confronts, and its contribution to the fusion innovation ecosystem, will be examined in Chap. 6.

2.2.3 Spherical Tokamak for Energy Production (STEP)

Emerging from the UKAEA, and based in an organization called UK Industrial Fusion Solutions (IFS), the Spherical Tokamak for Energy Production (STEP) aims by 2040 to produce a fully integrated fusion energy plant prototype capable of supplying electricity to the grid. The *UK's Modern Industrial Strategy* announced in 2025 includes reference to STEP as part of its ambition to become a Clean Energy manufacturing and innovation superpower (UK Government 2025), and in June

[11] Assuming 1 MW is the average power requirement for 2000 homes for an hour: https://www.ofgem.gov.uk/ofgem-publications/76160/13537-elecgenfactsfspdf. 500 MW powers a city the size of Leeds or Baltimore.

Credit © ITER Organization, http://www.iter.org/.

Fig. 2.5 ITER site. (Credit © ITER Organization, http://www.iter.org/)

2025 the fusion programme received a record investment from the UK government of £2.5 billion over 5 years to further its development. Differing from the research machines, JET and ITER, designing and building the plant requires specific interactions with industry. Its aim is to develop industrial capabilities and provide educational and training opportunities and stimulate innovation in engineering, construction, systems, and operations. A site has been selected for its location. It will include a few partners in an integrated delivery team: the UKAEA as science and technology partner and partners for engineering and construction, along with several systems partners to work with the integrated team on the design and delivery of the many complex systems involved. The next design phase involves developing capabilities in test and manufacture of systems and subsystems. The history and plans for STEP, and its role in a fusion innovation ecosystem, are analysed in Chap. 7. Fig. 2.6 provides an image envisioning STEP, noting its more “apple” than “doughnut” shape.

Credit: UKIFS

Fig. 2.6 STEP. (Credit: UKIFS)

2.3 Fusion Energy, Fusion Power, and Fusion Technology

Fusion energy has multiple potential uses, including beyond electricity production feeding the electricity grid. It also can provide direct power to industry in, for example, large-scale energy-intensive manufacturing process such as steel production. As the massive energy requirements for cloud computing is making technology companies such as Amazon and Google consider the use of fission-based nuclear power stations adjacent to their data factories, fusion may eventually provide a better option.

The development of fusion energy also relies upon advances in related and supportive technologies and can contribute to those advances. This is seen in fields such as robotics, new materials, and digital design. Direct uses of fusion-related technology is seen in the medical field, in industrial imaging, and in nuclear waste handling. For example, TAE technology, a small fusion company based in California, has created a subsidiary oncology company, TAE Life Sciences, to commercialize a neutron beam used to irradiate tumours in the head and neck. It is adapting its accelerator-based technology for a non-invasive, highly targeted treatment for the most difficult cancers, and has started human clinical trials, aiming first for the Chinese market (Chinese people suffer extensively from these cancers). Another fusion company, Tokamak Energy, has developed world-leading expertise in high-temperature superconducting magnets, with potential industrial applications in medicine, renewable energy, and propulsion. Building on UKAEA research at Culham, spin-off companies have been created in industrial controls and autonomous vehicles. The range and depth of expertise in Culham has led it to be named as the UK's first AI Growth Zone in the UK government's flagship programme to build AI infrastructure and increase investment in AI-enabled data centres.

The Tony Blair Institute lists some of the technologies that have improved due to fusion research, including: energy-efficient micro-plasma lighting, manufacture of silicon chips and semiconductors, aircraft engines coated in plasma-treated powder to enhance safety and efficiency, and MRI scanners that have improved through advances in superconducting magnets used in fusion.[12]

2.4 Why Fusion Energy Matters

Faith is placed in the pursuit of fusion energy primarily because of its potential to deal with climate change. It is needed to contribute to increasing demand for cleaner sources of sustainable energy that do not produce carbon and emit greenhouse gases. Fusion energy addresses two of the United Nation's Sustainable Development Goals in particular: the aims to "ensure access to affordable, reliable, sustainable and modern energy for all" and to "take urgent action to combat climate change and its impacts". Global energy demand is estimated to increase significantly over the next decades. 850 million people around the world still have no access to electricity.[13] It is evident that at the same time as increasing the supply of energy there is a need to massively reduce reliance on CO_2-producing energy technologies.[14] The amount of fossil fuels burnt to create electricity nearly doubled between 2000 and 2020. Reducing reliance will require a shift towards electrification in many traditional fossil-fuel energy systems, for example, in transport, unless alternative forms of energy such as hydrogen fuel can be developed quickly and cost-effectively.

A clean energy transition is needed to meet the 2016 Paris Agreement of the UN Framework Convention on Climate Change, confirmed by COP26 held in Glasgow in 2021, of holding the rise in global temperatures from pre-industrial levels to "well below 2 degrees Celsius … and pursuing efforts to limit [it] to 1.5 degrees Celsius". China has pledged to be carbon-neutral by 2060, and the UK is committed bring all greenhouse gas emissions to net zero by 2050.[15] Becoming net zero by 2050 will entail a massive shift in energy mix, estimated to involve the displacement of 11 million tons of oil equivalent, or alternatively the building of a nuclear energy

[12] https://institute.global/insights/tech-and-digitalisation/fusion-coming-and-its-bringing-more-clean-energy.

[13] https://www.iea.org/reports/world-energy-outlook-2019#.

[14] Global Energy Outlook 2020. *Becoming net zero*.

[15] https://www.scmp.com/news/china/diplomacy/article/3106509/climate-change-chinas-goal-be-carbon-neutral-2060-will-cost. Net zero involves balancing the amount of greenhouse gas emissions produced and the amount removed from the atomosphere.

https://www.gov.uk/government/news/uk-becomes-first-major-economy-to-pass-net-zero-emissions-law.

station or massive offshore wind farm every day for 30 years.[16] One estimate suggests that the world will need 12 times the amount of clean electricity than is produced today.[17]

While fusion energy is in development, there are major advances taking place in the delivery of renewable energy sources. Renewables made up over 25 percent of global energy generation by 2018, and this is expected to rise to 45 percent by 2040.[18] Most of the increase will likely come from solar, wind, and hydro-energy, which are becoming increasingly cheap sources of electricity. As Hirst (2018), from the Grantham Institute suggests, demands from such renewables will be increasingly important if electricity becomes the preferred low-carbon solution for heat and transport and in the absence of commercial carbon capture and storage. Renewables can meet a significant proportion of demand, but they cannot guarantee minimal requirement baseloads because of vagaries in the weather and may be less suited for high-density cities. Conventional fission-based nuclear energy can provide baseload, but it has its own attendant costs and risks to be mitigated. Nuclear energy provides the largest source of low-carbon electricity in the developed world and the second largest, after hydro, in the world as a whole (Hirst 2018). Fusion energy's advantage lies in the way it could buffer nature's volatility and the intermittency of such renewables, avoiding the challenges it faces in storage, providing a level of baseload without the difficulties associated with fission energy.

Industry, transport, and agriculture consumes far more energy through their direct use of fossil fuels, rather than through using generated electricity. Decarbonizing the majority of the world's energy mix will therefore require sources of energy production that can deliver direct power, through heat, or, create hydrogen as a store of energy to be used in, for example, shipping and heavy transport. Fusion offers the possibility of producing energy for direct power and hydrogen production without degrading the environment.

In summary, fusion essentially offers five potential advantages in meeting the world's energy needs compared with alternatives. First, it could be used to provide a constant source of clean electricity, sited anywhere in the world, day and night, whether the wind is blowing and Sun is shining or not. Second, it could be used to provide direct power for industry, with the opportunity to conveniently locate it close to industrial sites or energy-intensive AI/Cloud data centres. Third, it could be used to produce hydrogen, which would then be used to power transport, particularly shipping and perhaps in future, air travel. Fourth, it is the densest form of energy production: more can be produced in a smaller space. Fifth, it has lower risks in managing its waste compared to fission plants or impacting local ecologies compared to solar, wind, or tidal technologies.

---

[16] Ian Chapman, Kavli Lecture, Royal Society, 27th January, 2021.

[17] https://www.ted.com/talks/solomon_goldstein_rose_how_much_clean_electricity_do_we_really_need/transcript?language=en.

[18] https://www.c2es.org/content/renewable-energy/.

The European Atomic Energy Community, Euratom, estimates fusion energy could provide 10 percent of the world's energy, roughly the amount expended in Europe, around 1000GW, by the end of the century, although this could increase if capital costs are lowered.[19] The company Energy Monitor contends that fusion could contribute a valuable energy source in a post-carbon society beyond 2050.[20] The fusion community believes that development has moved beyond the experimental stage and is entering the delivery era and that it will become a major part of world energy supply from the second half of this century onwards.

In essence, the speed with which fusion becomes viable will depend on the resources dedicated to it, the intensification of innovation ecosystems that surround it, and the continuing productive efforts of the people who work on its development, industrialization, and commercialization.

References

I. Chapman, S. Cowley, H. Wilson, The Spherical Tokamak for Energy Production: theme issue introduction. Phil. Trans. R. Soc. **A 382**, 20230416 (2024)

D. Drollette, Introduction: Fusion, the next big thing—Again? Bull. At. Sci. **80**(6), 343–344 (2024)

N. Hirst, *The Role of Nuclear Electricity in a Low-Carbon World, Grantham Institute Briefing Paper 24* (Imperial College, London, 2018)

T. Hughes, *Networks of Power: Electrification in Western Society 1880-1930* (Johns Hopkins University Press, Baltimore, 1983)

M. Irvine, *Nuclear Power: A Very Short Introduction* (Oxford University Press, Oxford, 2011)

N. Plfeugers-Peters, From ignition to energy, in *Science & Technology Review*, (Lawrence Livermore National Laboratory, 2025)

A. Sakharov, *Memoirs* (Random House, New York, 1990)

A. Turrell, *The Star Builders: Nuclear Fusion & the Race to Power the Planet* (Weidenfeld & Nicolson, London, 2021)

J. Uglow, *In These Times: Living in Britain Through Napoleon's Wars, 1793–1815* (Faber and Faber, London, 2015)

UK Government, *The UK'S Modern Industrial Strategy* (HMSO, 2025)

[19] https://www.euro-fusion.org/eurofusion/roadmap/.

[20] https://twitter.com/energymonitorai/status/1488178800017952774?s=11.

Part II
People

Chapter 3
Standing on the Shoulders of Giants

The pursuit of fusion energy is garlanded with some of the greatest names in science and is a perfect illustration of Newton's statement that if he saw further it was because stood on the shoulders of giants. The story progresses from Marie Curie's nineteenth century discovery of heat emanating from radium, a radioactive material that was to contribute to killing her, to Ernest Rutherford's 1904 detection of the enormous energy released by radiation from radioactive substances, to Albert Einstein's famous 1905 formula $E = mc^2$ showing that a tiny amount of mass could be converted into a massive amount of energy. Curie remains the only scientist to have won the Nobel Prize for both Physics and Chemistry. Einstein is remembered as one of the greatest ever physicists. The extent of Rutherford's contribution led him to be known as the "father of nuclear physics", and to him being buried in London's Westminster Abbey close to Newton and Darwin.

The journey of discovery continued in the 1920s with F.W. Aston's measurements of the masses of different atoms, including hydrogen and helium. Aston's finding that hydrogen nuclei were heavier than a helium nucleus—that is, a helium nucleus is lighter than would be expected by combining two individual protons and two neutrons—was recognized the same year by Sir Arthur Eddington as the explanation that the Sun shines by converting hydrogen atoms to helium. Eddington, a British astronomer, physicist and mathematician, was an early advocate of Einstein's theory of relativity. At his Presidential address to the British Association in Cardiff in 1920, using Einstein's famous $E = mc^2$ formula, he explained how stars, such as the Sun, draw their endless energy from the fusion of hydrogen into helium. He went further in his speech to suggest that one day it might be possible to harness that energy on Earth:

> A star is drawing on some vast reservoir of energy by means unknown to us. This reservoir can scarcely be other than the sub-atomic energy which, it is known, exists abundantly in

M. Dodgson, D. Gann, *The Search for Viable Fusion Energy*, SpringerBriefs in Applied Sciences and Technology, https://doi.org/10.1007/978-3-032-13466-0_3

all matter; we sometimes dream that man will one day learn how to release it and use it for his service. The store is well-nigh inexhaustible, if only it could be tapped.[1]

Leo Szilard, the Hungarian-American physicist, held a 1934 patent on atomic energy that covered the principle of fusion[2] although proof that energy is liberated when light nuclei combine to form heavier ones, i.e. fusion, was not produced until 1939 by Hans Bethe at Cornell University. Einstein and Bethe both became prominent voices against the use of nuclear science in warfare, and it is testament to Eddington's foresight that as well as seeing the opportunities of the science, there was also risk:

> If, indeed, the sub-atomic energy in the stars is being freely used to maintain their great furnaces, it seems to bring a little nearer to fulfilment our dream of controlling this latent energy for the well-being of the human race—or for its suicide.[3]

It was Mark Oliphant, a student of Ernest Rutherford, at the Cavendish Laboratory at Cambridge University, who in the 1930s first experimentally demonstrated nuclear fusion. By firing heavy hydrogen nuclei in a particle accelerator, he discovered helium-3 and tritium, showing they could be made to react with each other, releasing energy.

Fusion, like so many scientific discoveries, was a truly international effort, and its history includes contributions from Curie (Polish-French), Einstein (German-American), Rutherford (New Zealand-British), Eddington (British), Szilard (Hungarian-American), Bethe (German-American), and Oliphant (Australian). And, like all scientific discoveries, these great scientists not only stood on the shoulders of other giants, but relied on the work and support of hundreds of less celebrated colleagues around the world. The story of the development of fusion energy is one of lengthy and deep international scientific collaboration.

A number of aspects of the early development of fusion energy, illustrating its scientific complexity, the importance of personal connections, and its extraordinary capacity to transcend political hostilities, are illustrated in the following story of a British physicist, Mike Forrest.

3.1 Hot Experiments in a Cold War[4]

What Forrest remembered first after landing at Moscow airport in December 1968 was how incredibly cold it was, at minus 30 degrees Celsius. The irony was his mission at the time was to record the hottest temperature yet achieved on earth, over ten million degrees Celsius.

---

[1] Arthur Eddington, Presidential Address to the British Association, Cardiff, August 14th, 1920. Published in *The Observatory*, 43: 341–358, 1920.

[2] Cited in Hendry and Lawson 1993.

[3] Quoted in Bahcall (2000).

[4] Sources include: British Library, Voices of Science, https://www.bl.uk/voices-of-science/interviewees/michael-forrest/audio/michael-forrest-first-day-in-the-ussr.; and Forrest, M. (2011); https://www.youtube.com/watch?v=-wJaFAtCKPc.

Forrest was part of a five-person British team that had been invited to conduct a nuclear experiment during the height of the Cold War political hostilities between the USSR and the West. Amongst all the fears of thermonuclear war, and "mutually assured destruction", scientists from Britain and the USSR were to work together to advance nuclear science.

A diagnostic physicist, Mike Forrest's job was to find a way of accurately measuring electron temperature, density, and confinement time, all of which are essential for scientific progress in fusion, especially in the plasmas in which fusion occurs. He began working on an early fusion machine called Zeta (Zero Energy Thermonuclear Assembly) at the UKAEA in 1957. The story of ZETA will be told in Chap. 5. Lack of precise data was holding back the research and in 1964 his attention was brought to a letter by Tom Hughes in *Nature* which described the possibility of using lasers to get a direct measurement of plasma temperatures and densities through "Thomson Scattering". Forrest quickly worked on the idea, developing his own experiments and later that year published a paper in *Nature* confirming results (DeSilva et al. 1964). As light scatters from plasma, it can be detected to accurately measure the plasma's density and temperature. This approach was thought to be potentially an important way of increasing accuracy of measurements, and Forrest was given an "essentially limitless" budget to develop the technique. His team was moved to a purpose-built laboratory at the UKAEA's Culham facility to test whether the technique would work in practice. The team had to build its own laser because the ones usually bought were not powerful enough. The project proved successful and was used in Culham's fusion research.

In 1968, the Soviet Union surprised the scientific world with the announcement that its Tokamak T3 device in the Kurchatov Institute in Moscow had produced temperatures of ten million degrees Celsius, ten times hotter than anything achieved previously. The tokamak design was torus-shaped with plasma contained by magnets and a powerful current running through the plasma to stabilize and heat it.[5] The Soviet claim was treated sceptically, especially in America, either for political reasons or because the results were so different from the experiences of US scientists. If the numbers were right, however, then the experiment had demonstrated that fusion was feasible; a huge fillip for the fusion research community.

The Director of the Kurchatov Institute, Lev Artsimovich, one of the USSR's most eminent scientists, and Andre Sakharov's boss, had met and befriended Bas Pease, the Director of UKAEA, at a Pugwash Conference. These conferences were designed to bring scientists together to promote world peace. Faced with scepticism about the tokamak's performance, Artsimovich asked Pease if a team could be sent from Culham to confirm the results by measuring electron temperatures and pressures.

Scientifically, politically, and personally this move was risky. Accepting the invitation implied the British admitting the Soviets had the lead in fusion technology, and the Soviets admitting they did not have the means to measure the temperature

[5] The principles of using magnetic confinement and radio frequency plasma heating in a toroidal vacuum chamber in nuclear fusion reactions was first recorded in a patent taken out by the UKAEA in 1946 based on the research of G. P. Thomson and M. Blackman (Irvine, 2011).

of plasmas. The risk to the Soviets was their measurements that were inaccurate and their claims were wrong, the risks to the British was their measurement techniques didn't work at such high temperatures. Mike Forrest was given a last-minute opportunity by Bas Pease to pull out of the project. When Forrest affirmed he was going ahead, Pease said that it was "very brave". Artsimovich was similarly bold. Despite his elevated status in the USSR, failure would have been scientifically, diplomatically, and personally very damaging. It was recognized, however, that confirmation of the tokamak's achievements, although risky, would bring benefits for the science and to both nations.

Preparing the equipment for the tests took 3 months, during which the team was given top priority access to all UKAEA resources. Five tons of equipment was flown to Moscow in March 1969 on a specially arranged Pakistan Airlines flight, the only airline that could take it. With approval from the British Foreign Office and Ministry of Technology, and an open mind about the Russian claims, the five British scientists began work in the enormously sensitive field of nuclear science in a highly hostile diplomatic environment behind the Iron Curtain.

Relations between the British and Soviet scientists, in contrast, were extremely friendly and productive. There were many opportunities for socializing and good friendships were made. Some KGB interest in the British scientists was noted, but not considered problematic.

The measurement project, however, faced enormous technical difficulties. Vibrations on the floor of the tokamak needed to be countered. It was found that the measurements of background light were incorrect; so much so that a new machine needed to be built. Prisms exploded and needed to be replaced. Time was pressing. With all the delays, Artsimovich was showing the pressure. Parts needed urgently were sent in diplomatic bags. The team was working all hours.

Through an extraordinary collaborative effort, in July 1969 the test was successful. Clear and widely spread scattered light was recorded above background light. It confirmed that temperatures of ten million degrees Celsius had been achieved. The main feeling in the team was relief.

The results of the tests were published by scientists from UKAEA and Kurchatov Institute in *Nature* in November, 1969 (Peacock et al. 1969). After Bas Pease learned of the confirmation test he telephoned his American counterpart, Harold Furth, the Fusion Director at Princeton University, to tell him. Princeton immediately changed its existing stellarator machine and created a tokamak within 4 months. Six months later it matched the T3 results.

What Mike Forrest called a "unique penetration of the Iron Curtain" demonstrated the continuing possibility of international scientific cooperation in the most unfavourable of political circumstances. Also importantly, British and Soviet technology had combined to confirm the achievability of fusion in a controlled experiment. As Forrest put it: "If we hadn't developed the Thomson Scattering measurement technique, fusion would have developed at a much, much slower pace". Science and the radical innovations that can spring from it often only makes progress because of breakthroughs in the ability to see and measure the world, and the history of science can in part be read through the lens of the development of scientific instruments

(Crump 2002). Plasma physics and the ability to create fusion are exemplars, relying on the development of Thomson Scattering measurement capabilities. One advance in one technology is often reliant upon progress in another. A direct line can be drawn between the proof of the tokamak concept in the USSR in 1969 to the construction of JET in the 1980s, ITER in the 1990s/2020s, and STEP in the late 2020s/2030s.

Scientists working on the pursuit to harness fusion are confident that it will be achieved, but rarely underestimate the profound challenges still to be confronted. They are highly aware of the scientific and engineering difficulties to be addressed before fusion energy is viable, and how extraordinarily hard it will be to realize it in practice. Hubris—excessive pride, or in Greek myth defiance of the gods—is most unwise in the face of such daunting challenges. Like Helios's journey, distractions and surprises can delay and subvert progress. And scientists, being human, make mistakes.

3.2 When Modesty Fails: The Cold Fusion Story

On March 23rd, 1989, Martin Fleischmann, Professor of Electrochemistry of the University of Southampton, UK, and Stanley Pons, chairman of the chemistry department at the University of Utah, USA described an experiment in general terms at a press conference in Salt Lake City. They claimed their experiment had produced fusion at room temperature. In so doing, they began not only the most controversial episodes in fusion research, but one of the most controversial affairs in the whole history of science.

In what followed were years of divisive arguments over priority in discovery (who discovered it first), secrecy around potentially valuable intellectual property, and the replicability of experimental results, overlaid with the spectre of fraudulence. Chemists were pitted against physicists; proponents of hot fusion against cold fusion; and the scientific establishment was castigated for its resolute opposition to disruptive new thinking. Millions of dollars were invested by universities and companies in the search for cold fusion, almost all fruitlessly.

Scientists are sceptics. The motto of the Royal Society, one of the world's most prestigious scientific academies, is *nullius in verba*, which essentially means "take nobody's word for it". The announcement by Fleischmann, who was himself a Fellow of the Royal Society, and Pons was immediately treated with scepticism, not least because the scientists chose to announce their findings in a press conference, rather than by the usual method of peer-reviewed publication. The contrast was stark, furthermore, between the accepted wisdom that fusion occurs at extremely high temperatures and the claim it was found at room temperature. Fusion experiments were associated with large, complex scientific instruments; Fleischmann and Pons' experiment applied a current to a cell, believing deuterium atoms from heavy water had penetrated into a palladium cathode fusing to form helium atoms, with the excess energy from the process dissipated as heat. Their table top instrument essentially consisted of a glass beaker in a bowl.

Much has been written about this affair, its claims and counter-claims (see, for example, Siefe 2008; Ball 2019). The scientific community has dismissed the experiment as poorly constructed and its claims as unsustainable and exaggerated. Cold fusion has been called "pariah" and "pathological" science, and it has attracted its share of "complete cranks" (Siefe 2008). Yet cold fusion research continues around the world, rebranded as low-energy nuclear reactions—LENR (Ritter 2016). As well as an indictment of the motives and behaviours of individual scientists, it can also be seen as an indication of the extraordinary latency in the demand for fusion energy. At the notorious 1989 press conference, the President of the University of Utah, a medical researcher, claimed the discovery of cold fusion could be compared to the discovery of fire and electricity.

Although the science of cold fusion was unproven, the idea captured the imagination of even supposedly technology savvy investors. Neil Woodford, for example, a venture funder with a reputation at the time for being one of the UK's best investment managers, invested around £54 million in Industrial Heat, a North Carolina company attempting to create energy from low-energy nuclear reactions.[6] It was the biggest holding of Woodford Patient Capital Trust, which had a reputation for bold investments. This fund ceased trading in 2019, shrouded in controversy and the subject of Financial Conduct Authority suspension and investigation. Other investors in Industrial Heat included the actor Brad Pitt and Laurene Powell Jobs, wife of the late Apple founder, Steve Jobs. In 2016, amidst legal claim and counter claim, the company ended its relationship with the person who sold the firm his claimed intellectual property in cold fusion technology.[7]

Toyota is reputed to have invested $40 million in cold fusion research.

Google was also attracted to cold fusion, investing $ten million from 2015 in a 4-year programme with a team of 30 researchers experimenting with low temperature nuclear reactions. Results from Google's research were published in *Nature Perspectives* in 2019, reporting no evidence of the phenomenon.[8] The team did claim advances in materials science and chemistry which may be of use for other applications such as hydrogen storage.[9]

As well as the risks of investing in unproven and emerging research, cold fusion also highlights some of the strengths and weaknesses in the current conduct of science. It starkly reveals the pressures to demonstrate priority in publishing and patenting, which can lead to some unseemly interpersonal and inter-organizational conflicts. The search for personal prestige and commercial opportunities are powerful motivators. The case shows the strengths of the peer review system. Fleischmann and Pons submitted a paper on their research to the leading scientific journal, *Nature*, but never responded to comments by its referees. It also shows the immense

[6] https://www.ft.com/content/024cfc4a-8df6-11e9-a1c1-51bf8f989972.

[7] https://news.newenergytimes.net/2019/06/07/bad-cold-fusion-investment-contributes-to-woodford-fund-collapse/.

[8] https://www.nature.com/articles/d41586-019-01683-9.

[9] https://www.technologyreview.com/2019/05/29/135179/google-has-been-funding-research-into-cold-fusion-for-years/.

difficulties in replicating scientific experiments: highly regarded laboratories initially reported support for cold fusion phenomena before recanting. It further reveals the extent of the opposition from entrenched interests to radical and disruptive ideas. Perhaps most revealing is the importance in science in avoiding hype. Had Fleischmann and Pons' experiment been reported as an interesting, unexplained phenomena, rather than claiming the discovery of a new, cheap, and plentiful energy source, then the journey of those searching for such a source would not have been tarred by such a damaging diversion.

3.3 Curiosity

Curiosity, the human desire to understand the wonders of the world, is a powerful, less instrumental, motivation for the pursuit of fusion energy. At its extremity, in this view, fusion energy should be pursued simply because it is hard: it profoundly stretches the imagination and challenges the practical. President John F. Kennedy's famous speech on September 12th, 1962 has the memorable line about why humans aspire to go to the moon and solve other technological challenges, "not because they are easy, but because they are hard" (Kennedy 1962). He cites William Bradford, a Puritan who sailed on the Mayflower, who wrote in 1630 that "all great and honourable actions are accompanied with great difficulties, and both must be enterprised and overcome with answerable courage". To the virtues of honour, initiative and courage, Kennedy adds determination, choice, knowledge, and understanding. He refers to enrichment of new knowledge, new techniques of learning, mapping and observation, and new tools and computers. Kennedy's speech captures the essential inspiration that underlines the connection of human ambition with scientific and technological possibilities and is manifested in the search for fusion energy. The pursuit of challenging scientific and technological problems is one of humankind's most noble characteristics.

Curiosity is an intrinsic feature of humanity, a key motivator for many scientists and engineers, and a major stimulus to the accumulation of knowledge. The thirst to understand lies behind humankind's urge to conquer flight, to go to the moon, and to discover the structure and function of DNA. Feeding our theme of a pursuit or quest, curiosity is endless and constantly renewable. It is fed by the ability to see, notice, question, and seek answers. Curiosity is defined by the Oxford English Dictionary as "the desire or inclination to know or learn about anything, especially what is novel or strange". Hans Bethe said mankind has a fundamental urge to comprehend the world, a compulsion to understand. For neuroscientists, curiosity is a basic and pervasive element of human nature (Kidd and Hayden 2015). And it can lead to great personal rewards; as the great scientist Linus Pauling said: "satisfaction of one's curiosity is one of the greatest sources of happiness in life".[10]

[10] Quotations from the *Oxford Dictionary of Quotations*, eighth edition, Oxford, Oxford University Press, 2014.

This desire, urge, or compulsion has long attracted the attention of philosophers; lauded by some, distained by others. Curiosity has been associated with being nosy, having interests beyond your station in life and, of course, it killed the cat. Aristotle preferred the term wonder, and St Augustine thought curiosity a disease. From Sir Francis Bacon in the sixteenth and seventeenth century to Michel Foucault in the twentieth century the nature and importance of curiosity has featured in philosophical thinking about the human condition.

For the arch-monarchist Bacon, the curiosity-driven pursuit of knowledge allowed control over the world to further the power of the king. Science, in his view, was to be the servant of the state. Modern philosophers look to more intrinsic motives for curiosity. For Foucault (1980) it:

> evokes "concern"; it evokes the care one takes for what exists and could exist; an acute sense of the real which, however, never becomes fixed; a readiness to find our surroundings strange and singular; a certain relentlessness in ridding ourselves of our familiarities and looking at things otherwise; a passion for seizing what is happening now and what is passing away; a lack of respect for traditional hierarchies of the important and the essential.

Francis Bacon was aware of the downsides of curiosity, especially in the vanity of knowledge pursued for the wrong reason, warning that it is used "seldom sincerely to give a true account of their gift of reason, to the benefit and use of men". In his 1605, Treatise on *The Proficience and Advancement of Learning*, he argued that curiosity arose from:

> a desire of learning and knowledge, sometimes upon a natural curiosity and inquisitive appetite; sometimes to entertain their minds with variety and delight; sometimes for ornament and reputation; and sometimes to enable them to victory of wit and contradiction; and most times for lucre and profession. (Bacon 1605).

Bacon's insights resonate strongly with today's psychologists who explore curiosity and motivation. Silvia (2019), for example, suggests three strands of thinking in the psychology of curiosity, it overcomes the unpleasantness of uncertainty, ignorance, or information gaps, akin to scratching a mental itch; is a source of intrinsic motivation that fosters learning and exploring for their own sakes; and, is a source of difference from and advantage over the incurious, leading to higher academic achievements and sense of meaningful lives.

It certainty motivates scientists: Robert Pennock places curiosity first and forefront in his research on scientists, arguing:

> honesty and curiosity are the core values of scientific practice, and it is curiosity that is the most characteristic of the scientific virtues, and all other scientific virtues take on their distinctive character by virtues of their relationship with it. (Pennock 2019:xxvii).

And it is similarly a motive for engineers: Apple's Steve Jobs often drew the connection between curiosity, creativity, and innovation.[11] In 2012, a remote controlled vehicle sent to Mars, involving the most advanced engineering of a machine ever sent to another planet, was named Curiosity. In a cynical but amusing observation

[11] See, for example, Turak, A., "Steve Jobs and the one trait all innovation leaders share", Forbes, November 21st, 2011.

about fusion, Lev Artsimovich, who played such a central role in the early development of fusion in the USSR observed: "Science is the best way of satisfying the curiosity of individuals at government expense".[12]

Curiosity is rarely welcomed by institutions and organizations prioritizing order over questioning and change. Andrei Sakharov's criticisms of the Soviet Union led to his exile and years of hardship, and he offers wise insights into the personalities needed to persevere with the application of science and for the development of the tokamak:

> On the face of it the situation was almost precisely that of Einstein's famous description of how inventions occur. First, the experts present weighty arguments why it's impossible; then an ignoramus appears who doesn't know any of this – and goes ahead and makes the invention. This must be taken with a grain of salt: the so-called ignoramus must in fact be up to the state of the art and possess certain qualities, or he won't get anywhere. He should be aware of the difficulties but have intuition enough not to be put off by them, even if he can't yet provide a strictly logical case for his belief. (Sakarov, 1990:143)

These human traits in scientists and engineers has long manifested itself in groups and organizations marshalling and protecting their curiosity, such as Gresham College and the Royal Society in the sixteenth century and the Lunar Society in the eighteenth century (Uglow 2003). Curiosity has also been furthered by the development of instruments and machines. Instruments have advanced science and engineering for centuries, with Galileo's telescope being the most famous example, and we have seen the importance of lasers measuring the temperature of plasmas. When Thomas Sprat wrote his *History of the Royal Society* in 1667, it included an illustration that included a bust of Charles II and a portrait of Francis Bacon. In the background is an air-pump, owned by Robert Boyle, that Ball (2012) describes as the most sophisticated scientific instrument of the seventeenth century. Ball makes the connection between the air-pump with today's machines for particle physics, both deliver vacated space, and both encourage and enable experimental curiosity. He also writes how relatively simple an instrument to contemporary eyes, proved difficult to use, pushing the boundaries of materials and engineering knowledge of the time. Ball also cites a former Director of CERN explaining the presence of the Large Hadron Collider by referring to the role of human curiosity.

That is not to say that curiosity about fusion can only exist in large-scale experiments. Fusion reactions have been created by a wide range of individuals and laboratories around the world, including most remarkably by a 15-year-old boy (Clynes 2015).

All these aspects can be seen amongst the scientists and engineers marshalling their curiosity to work on fusion today in organizations such as the UKAEA, ITER, and other institutions around the world. It can involve a range of motivations, from the most noble and altruistic—such as solving the climate crisis—to the most instrumental—Bacon's lucre, such as seeking profit and career advancement—but amongst all the driving forces behind the development of fusion energy—the pressing need, and all its resources and equipment, politics and organization—is the

[12] *Oxford Dictionary of Quotations*, eighth edition, Oxford, Oxford, Oxford University Press, 2014.

innate and essentially virtuous characteristic of intellectual challenge. Einstein said he was driven by passionate curiosity. Whether it is directed at the most basic of scientific questions—what Einstein called the holy curiosity of inquiry—or to the challenges faced by engineers—who Einstein called the architects of progress—the delivery of fusion energy is driven by curiosity. Although there has been enormous progress towards the delivery of fusion energy, scientific, technological, economic, and social questions remain. What is unquestionable is the pursuit of fusion energy is a quest constantly fed by curiosity; and curiosity, of course, is not just the preserve of science.

3.4 Two Cultures

Profound human endeavours, such as the pursuit of fusion energy, are better understood when their purpose and significance are connected to as broad a perspective as possible. CP Snow (1961) famously wrote about the two cultures of science and the arts, and the need for better connections between them for the advance of knowledge and social progress. Incredible insights emerge when visionary and imaginative authors explain great science and scientists (see, for example, Labatut 2020, 2023). Curiosity in understanding the world does not obey disciplinary boundaries, and bringing the process that heats the Sun to Earth cannot be appreciated only as a scientific endeavour. We claimed at the start of this book that the story of fusion energy requires both prose and poetry. For CP Snow, changing the world for the better depends on the wisdom that derives from the unification of knowledge and understanding of science and the arts. The ambition to replicate something so fundamental as the source of all life, capturing something so vital and elemental to our experience of the world, in literature, painting, music, and religion, means that alongside the prose of engineering and technology, the pursuit of fusion energy also needs to be embraced and explained by the arts.

We have seen the fascination of poets with machines, how popular authors excite the imagination of the public about machines, and how philosophers explain why scientists are curious. And this connection of the arts and science is especially vivid when it comes to the Sun. The Sun is everywhere in our understanding of the world—and has been, as we have seen, a core feature of our mythology—and to the effort to try to create a Sun on Earth has deep cultural meaning and says something profound to humankind.

The Sun has proverbially been associated with brightness and clearness, a source of glory, inspiration, or understanding (Knowles 2005). Capturing the significance and meaning of the Sun has been a major motivation of artists. Great paintings of the Sun, such as by Monet, Turner, and Van Gogh, capture its splendour and how all that is beautiful depends upon it. Shakespeare's King Lear speaks of the "sacred radiance of the Sun", and in Romeo and Juliet asks: "…what light through yonder window breaks? It is the east, and Juliet is the Sun". In Coleridge's Rime of the Ancient Mariner, one of the greatest poems in the English Language, are the lines:

"Nor dim nor red, like God's own head, The glorious Sun uprist". Coleridge's connection of the Sun with God has a long history beyond classical mythology: Egyptians, Inca and Aztec, Persians and Celts, are joined by Hindus and Buddhists in the deification of the Sun. Whether known as Ra, Mithras, Surya, or Huitzilopochtli, the Sun is a god or god-like.

In music, there are Hayden's *Sun quartets* and Nielsen's *Helios Overture*, and the movement of the Sun is a central feature of Grieg's *Peer Gynt* and Strauss's *Also sprach Zarathustra* and *Alpine Symphony*. More contemporarily we have Sally Beamish's *The Imagined Sound of Sun on Stone*, Max Richter's *On the Nature of Daylight*, and Rihanna in her album *Music of the Sun*, sings: "Doesn't matter who you are or where you're from. Come and dance to the music of the Sun". From high-brow to popular culture, the Sun remains a focus of attention as being core to human experience.

Art and science share the virtues of curiosity, creativity, and imagination and can inspire each in the other. The connection between the art of the Sun and of its science is perhaps best captured in Edvard Munch's masterpiece, *The Sun*, which portrays both its energy-giving force as the essence of life and creativity and reference to scientific breakthroughs at the time in X-rays and magnetism.

In the end, if we want to explain something so profound as the ambition to emulate in a machine the forces that create and support all life, we also need poetry. It takes a scientist to create a plasma, but an artist to capture and reveal the beauty of its vivid pink and fuchsia colours and swirling flows. The effort and impact of the science and technology of fusion energy assumes even greater consequence when we also account for the perspectives of the arts in the ambition of bringing the Sun to Earth.

Beyond the wider sense of science and the arts, cultural differences are also strongly evident nationally and organizationally and within the innovation process itself. National cultural differences bringing different perspectives and practices are valued in large scientific experiments, such as CERN and the Square Kilometre Array Telescope, and organizational cultures have long been known to vary between those elements concerned with exploring new avenues of expertise and those involved with exploiting existing ones. Cultures vary within and between firms in areas such as their approach to purpose, tolerance of risk, capability building, and attitudes to employees. The term "R&D" includes basic research, unfettered by consideration of its use, and industrial development, concerned primarily with applications: each supported by very different cultures, encouraging of discovery on the one hand, and profit on the other.

The development of fusion highlights the general question around the measurement and quantification of discovery, risk, and speculation, and how common metrics of performance limit ambition. A risk-taking culture is often incompatible with systems driven by key performance indicators, stage-gate decision processes, and traditional financial measures such as return on investment. The challenge for leaders in fusion is to comply with the disciplined requirements of investment decision-making and accountability bodies, while protecting organizational cultures from the constraints that these impose on freedom, imagination, and serendipitous discoveries.

3.5 Leadership

As fusion has developed as a science and has increasingly integrated into an innovation ecosystem, it has required leaders with significant scientific credentials who are also capable of working effectively with government and industry. Examples include Professor Sir Steven Cowley, previously CEO of the UKAEA and Director of the Princeton Plasma Physics Laboratory (PPPL) since 2018. He is a Fellow of the Royal Society and has held academic posts at Imperial College and UCLA, and was President of Corpus Christi College at Oxford University. A theoretical physicist, he is an authority on astrophysical plasmas and nuclear fusion, and he has stated how important it is to increase PPPL's connections with business. His successor at UKAEA, Professor Sir Ian Chapman, is a Fellow of both the Royal Society and the Royal Academy of Engineering, and was CEO of the UKAEA between 2016 and 2025, overseeing a fundamental transformation of the organization (described in Chap. 6). An award-winning scientist, with deep connections to the international fusion research community in the USA, Europe, and China, Ian Chapman is also a talented communicator.[13] His leadership skills are reflected in his appointment in 2025 as CEO of UK Research and Innovation, the organization that oversees the UK's seven research councils with a £8.8 billion budget.

A different skill set is required from leading scientific research institutes to leading the development of commercial prototypes, and the UKFIS/STEP programme is headed by Paul Methven. With a background in the British Navy, where he had led on active service, he was a Rear Admiral and Director of Submarine Acquisition, including leading the Dreadnought submarine project with a budget of around £40 billion. Initially appointed as a Director of UKAEA, he assembled a small team to start designing a STEP as a prototype for electricity production. Beyond the design of the machine and its funding, decisions were needed on its regulation, location, and its most appropriate institutional and organizational structure (these issues are discussed in Chap. 7). Asked about what in his background aided him in his role, he answered:

> 33 years in the Royal Navy, and leading increasingly complex major programmes in defence. Learning to manage continual organizational evolution and change in the programme as you go along. Later in my career, the business of having to deal with multiple different facets of a problem. I'm very used to programme complexity in a government context, and while I'm not able to know precisely what's going to happen in this programme, I've got enough of a frame of reference to be able to understand the things that matter and the things that don't, and to give some degree of direction and confidence to the team that, don't worry, this will be okay, or actually we really do need to worry about this. So I think just the experience of playing the match several times before. But if I go way, way

[13] See, for example, https://www.youtube.com/watch?v=jPRlFAcuGec.

> back, actually, I think that if you get genuinely meaningful levels of responsibility very early in your career you start to develop the ability to take decisions and see them through. And you are comfortable with wearing personal risk

And this will require managing internally and with external stakeholders:

> One of the slogans that we use in this programme is that the boldest endeavours start with a practical step and what I tend to say in talking to the team is that success requires matching vaulting ambition and hard headed pragmatism. We need to be simultaneously be really ambitious and everything that comes with it such as being innovative, entrepreneurial, and pushing really hard. But equally, if we're just that, we're not going to deliver, and we have to equally be practical and pragmatic

> The riskiest position is the status quo ante. But government measures what you're doing today as a zero baseline for risk. Understandably they don't want government contracts to go wrong or waste public sector money. So it's necessary to frame and persuade and engage stakeholders in a way that they don't see things as risky. You spend quite a lot of time working at how to insulate what they perceive as risk, and see it instead as an opportunity. Rather than looking through the lens of what could go wrong, what we're interested in is what could go right, and how do we exploit that? So I think you have to have a really acute awareness and sensitivity to the perception of those stakeholders

The quality of leadership will be a key determinant in the future of fusion energy, an issue we'll return to in Chap. 8, and within fusion innovation ecosystems a critical skill will be to "boundary span", that is to work effectively across different kinds of organization. Such skills extend beyond executive teams within fusion organizations, and crucially include leadership within the Boards and Advisory Boards of both the public and private sectors and which are core components of their collective governance. Boards and senior leadership teams bring the essential domain expertise and personal networks needed in the collective leadership necessary as fusion innovation ecosystems evolve: as scientific organizations adopt more commercial perspectives, as private firms collaborate with research institutions, and as both connect with governments. The appointment of Board members and senior executives experienced with organizational transformation and the flexible mindsets to guide organizations as they rapidly evolve is and will be a defining element of successful leadership in fusion energy.

Unsurprisingly, as government has been the major player in fusion innovation ecosystems to-date, leadership in government has been, is, and will be fundamental to the pursuit of fusion energy. In the UK, Professor Sir Patrick Vallance, who was Chief Scientific Advisor to the government before becoming Minister for Science, has called for more scientific expertise in government, a conclusion echoed in a 2019 government review of science capability in government.[14] Governments might be assumed to struggle with first-in-the-world projects, such as STEP, where costs are high and there are no equivalent projects to calibrate them alongside. In projects, such as STEP, government using a traditional linear decision-making process might insist on assessing detailed designs before approving funding for industrial partners despite the particular expertise in those partners being necessary for those designs.

---

[14] (https://www.theguardian.com/politics/2020/jan/12/whitehall-needs-more-scientists-to-compete-with-china-patrick-vallance-dominic-cummings).

Yet alongside the leadership of Prime Ministers, fusion in the UK has benefitted significantly from the support and guidance of Ministers in Science, Energy, and Industry, and from public servants. The UK government has a Fusion Policy Team which has continually and consistently developed its knowledge about the potential and challenges of fusion over many years. It has been involved in negotiating relationships with EU partners during and after the disruptions of Brexit, and latterly with other Departments, such as the Prime Minister's Office 10 Downing Street, Treasury, and Cabinet Office, without which fusion would not have received its funding investment given the UK's tight budgetary constraints. Support for fusion energy amongst the UK public service can be argued to display the civic virtues of cautious, considered decision-making that transcends partisan politics and is described by Michael Lewis (2025) as being so indispensable for modern societies.

3.6 Careers and Commitment

When asked about his career highlights prior to his 2021 retirement ending a lengthy career working on fusion, including as the Head of JET, Lorne Horton, said it was undoubtedly the international community of scientists he had worked with. He had enjoyed working on different and challenging problems, but the greatest pleasure came from the friendships developed when working with others. He remembered the early years of JET, where he said there were teams defined by nationality and how it had changed over time into teams of Europeans defined by functional or scientific roles.

Scientific careers are often associated with words such as autonomy, challenge, development, meaning, variety, and collegiality. Their rewards are intrinsic rather than the extrinsic rewards of salaries. Such intrinsic rewards, however, are deeply affected by the circumstances in which people work.

Hackett et al. (2004) posit the idea of a "research ensemble", a group that connects research technologies, such as materials, methods and instruments, with others in the field and with politicians and policymakers. The latter groups participate in the construction and direction of the ensemble and influence its performance and the work of its members. The point is that researchers have very powerful attachments to their work, but these are strongly influenced by the state of the science and technology they are working on and external factors, such as government policies and the public acceptance of what they do. Hackett et al. remind us of Max Weber's view that the vocation of science entails both personal rewards and significant perils. Their study of two fusion research facilities of similar sizes, however, found very different levels of satisfaction with resources and, to lesser extents, with the intrinsic and extrinsic rewards of work. Some of the terms used by researchers in one fusion research centre included "embarrassment" (at not meeting expectations), "there is a lot of stress", "there is a lot of fear", and "we all feel we're being asked to do too many things too quickly".

Policy decisions affect the work of researchers in more ways than simply decisions over funding levels. Hackett et al. found that the demands on fusion researchers to operate devices above their design limits meant devices needed to be upgraded with long periods of downtime, which caused difficulties. They also found that in the USA, as seen in the case of PPPL, the story of which we will tell in Chap. 6, fusion research was unsettled and characterized by the abrupt funding and defunding of particular activities, including entire facilities and those they employ. They point out how the MFTF-B device at Lawrence Livermore Laboratory was mothballed on the very day its construction was completed. This is also seen in the US's position on ITER: leaving then re-joining. The authors also note that on occasion the US Congress has made decisions not only on funding levels but compelling substantive and technological changes in the national fusion energy programme without recourse to technical experts. In the past, fusion research moved politically from a "guidestar in the firmament (to) a fading landmark in contested terrain". As we shall see in Chap. 7, fusion returned back on the agenda in the USA, but still faces the uncertainties around scientific research in that country.

The implications are the need for a productive workforce in a settled policy context, involving continual and detailed information exchange with politicians and policymakers. In the context of fusion, this dialogue involves highly skilled and articulate advocates for fusion navigating between championing its potential and advances whilst acknowledging its challenges. As we have seen, leaders need to balance promise with delivery, creating a "buzz" that attracts attentions and resources whilst being realistic. In addition to the crucial role of such leaders, communicating with the public about fusion in a transparent and appropriately informed manner is essential. Both ITER and the UKAEA have extensive communications efforts, using a variety of platforms to convey their ambitions and achievements.

As alluded to by Max Weber, the motivations of scientists and technologists are often multiple and not usually instrumental. People working on fusion are highly motivated by the nature of their work and the ways they work determine the success of their enterprise. Bas Pearce, Director of Culham Laboratory 1967–1981, says of the development of science that "much of the interest lies in the inner history of human interactions in the highly technical environment", and that "scientists allow the undoubted pressure of events, as well as human emotion, to affect decisions".[15] Surveys of employees at the UKAEA reveal that many staff are aware that their knowledge and skills could lead to better paid jobs in parts of the country that are less expensive, but choose to work on fusion because of their belief in its importance. As one scientist said: "I want to leave a legacy". That is not to say that scientists have no concerns for pay and conditions. Tensions arose amongst those working on JET in the 1990s as European and British scientists working on the project had different employment contracts.

In film of scientists during the 1997 JET world record experiment (named DTE1) there is the sense of tension, pressure in overcoming problems on the day, relief at the outcome, and pride in the achievement. The scale of the accomplishment is

[15] Quoted in Foreword to Hendry and Lawson (1993).

described by a leader of the experiment, Fernanda Rimini, as "like a moonlanding". Rimini refers to the massive technical integration problems in conducting JET experiments but also the human integration challenges of having the right people with the right skills.: "everything you do has a person behind it". Still working on JET during the 2021 experiment (DTE2), she again captures the excitement by saying "you need to be an adrenaline junkie. You're doing something that has never been done before, never seen before…".[16]

The development of fusion energy relies, of course, on many more skills than those of scientists. Experiments wouldn't work, for example, without electricians skilled at using very high power. And the background of some people working on fusion are unconventional. As well as having technical degrees, Amanda Quadling, head of materials at Culham, has degrees in housing and third world development. It also involves the demands of operations and delivery. The roles of Lyanne Maclean, UKAEA's one-time Chief Operating Officer, for example, included ensuring everything was in place for operations—people, facilities, power, quality and safety controls, IT and engineering services, project and programme governance, and crisis management—and also delivering on the EU's JET contract. She was also responsible for managing the response to Covid-19, ensuring the facilities were maintained safely. The campus was closed for 6 weeks during the first lockdown during the pandemic, but 150 people kept essential equipment running and a strong emphasis on staff well-being and putting Covid protection in place. She said one of her greatest challenges was managing the growth in employee numbers, growing from 800 in 2017 to 2000 in 2021. Her planning skills were developed in a previous role as a Colonel in the British Army.

Much has changed: reports on the development of nuclear research during the 1950s and 1960s hardly mention any female participation. Lorna Arnold, a rare female employee of the UKAEA at the time reports on the concerns regarding the risks to women of being exposed to radiation in laboratories. One suggestion at the time was that only "unmarried ladies" should be employed in radiation areas, before it was decided to exclude all "women of reproductive capacity", regardless of marital status (Arnold 2012). In 2022, a "Women in Fusion" was created as a joint initiative of several organizations in nuclear research and engineering to achieve gender equality in nuclear sciences. In working environments such as fusion, the mission and challenging work attracts highly talented women who are often under-represented in senior roles elsewhere.

Revealing the importance of building a supportive culture, one of the first things that UKIFS did upon its formation was to create a charter on the culture, values, and behaviour demanded in the delivery of STEP. As well as emphasizing matters such as safety, it addresses issues such as well-being and inclusivity. This approach drew on the lessons from a number of major projects in the UK, and leaders of UKIFS

[16] *The Starmakers*. UKAEA film produced and directed by The Village Films. 2022. The film captures the immense difficulties faced in conducting JET experiments from a human perspective, its tensions and rewards.

stressed that everyone in the organization, and those working with it, such as suppliers, needed to subscribe to these values.

References

L. Arnold, *My Short Century: Memoirs of an Accidental Nuclear Historian* (Cumnor Hill Books, Palo Alto, 2012)

F. Bacon, *Of the Proficience and Advancement of Learning*, vol I (1605), pp. 10–11

J. Bahcall "How the Sunshines", (2000). https://www.nobelprize.org/prizes/themes/how-the- Sun-shines-2

P. Ball, Lessons from cold fusion, 30 years on. Nature **569**, 601 (2019)

T. Clynes, *The Boy Who Played with Fusion* (Houghton Mifflin Harcourt, 2015)

T. Crump, *Brief History of Science: Through the Development of Scientific Instruments* (Constable and Robinson, London, 2002)

A. DeSilva, D. Evans, M. Forrest, Observations of Thomson and co-operative scattering of ruby laser light by a plasma. Nature **203**, 1321–1322 (1964)

M. Forrest, *Lasers across the Cherry Orchard* (Stanford Publishing Ltd, Stanford, 2011)

Foucault, M. 1980. "The masked philosopher" in *Foucault live collected interviews, 1961–1984,*. edited by Sylvere Lotringer. 1996.New York, Semiotext (e):302-307

E. Hackett, D. Conz, J. Parker, J. Bashford, S. DeLay, Tokamaks and turbulence: Research ensembles, policy and technoscientific work. Res. Policy **33**, 747–767 (2004)

J. Hendry, J. Lawson, *Fusion Research in the UK 1945–1960* (Harwell, AEA Technology, 1993)

J. Kennedy, *Moon Speech* (Rice Stadium, 1962)

C. Kidd, B. Hayden, The psychology and neuroscience of curiosity. Neuron **4**, 449–460 (2015)

E. Knowles, *Oxford Dictionary of Phase and Fable* (Oxford University Press, Oxford, 2005)

Labatut, B., When We Cease to Understand the World, 2020. London, Pushkin Press

B. Labatut, *The Maniac* (Pushkin Press, London, 2023)

M. Lewis, *Who Is Government? The Untold Story of Public Service* (Allen Lane, London, 2025)

N. Peacock, D. Robinson, M. Forrest, P. Wilcock, V. Sannikov, Measurement of the electron temperature by Thomson scattering in tokamak T3. Nature **224**, 488–490 (1969)

Pennock, R., An Instinct for the Truth: Curiosity and the Moral Character of Science, 2019. Cambridge, MIT Press

S. Ritter, Cold fusion lives on. C&EN Global Enterprise **94**(44), 34–39 (2016)

C. Siefe, *Sun in a Bottle: The Strange History of Fusion and the Science of Wishful Thinking* (Penguin, New York, 2008)

P. Silvia, Curiosity and motivation, in *The Oxford Handbook of Human Motivation, second edition*, ed. by R. Ryan, (Oxford University Press, Oxford, 2019), pp. 157–166

C.P. Snow, *The Two Cultures and the Scientific Revolution the Rede Lecture, 1951* (Cambridge University Press, New York, 1961)

J. Uglow, *The Lunar Men* (Faber and Faber, London, 2003)

Chapter 4
Public Acceptance

Known particularly for his science fiction, HG Wells predicted the development of technologies that included lasers and email.[1] He also wrote about the threats and opportunities of nuclear energy. Writing in 1914 he said knowledge of nuclear physics could bring about "a change in human conditions that I can only compare to the discovery of fire", and its use could provide "a source of energy so potent that a man might carry in his hand the energy to light a city for a year, fight a fleet of battleships, or drive one of our giant liners across the Atlantic" (Wells 1914). He was also highly aware of its dangers, and his book *The World Set Free*, tells of the destructive power of atomic energy. The only way to avoid a dystopian future ruined by nuclear war, he argued, is unprecedented levels of international political collaboration and "the acceptance of achieved science as the basis of a new social order". Long before its harnessing in actual experiments, the massive transformational potential of the science behind atomic energy had reached the popular imagination.

HG Wells reminds us of the potential dangers of scientific discoveries which, as well as leading to marvellous improvements in areas such as health, have also given us weapons of mass destruction, the disastrous consequences of financial derivatives and securitization, and, of course, carbon-producing technologies. There is a large literature on the public acceptance of science and innovation,[2] and how new technologies become adopted and diffused is often complicated and not straightforward (see, for example, Ozaki and Dodgson 2014). Technology and innovations are, however, generally found to be more successfully accepted when they accord with the values, social goals, and concerns of the people adopting them. Adopters also have concerns about who benefits from innovations, and whether those bodies advancing them, be they researchers, private companies, or governments and

[1] https://www.smithsonianmag.com/arts-culture/many-futuristic-predictions-hg-wells-came-true-180960546/.

[2] https://ncses.nsf.gov/pubs/nsb20244/public-perceptions-of-science-and-technology.

M. Dodgson, D. Gann, *The Search for Viable Fusion Energy*, SpringerBriefs in Applied Sciences and Technology, https://doi.org/10.1007/978-3-032-13466-0_4

regulatory authorities, are trusted. The trust between citizens and these bodies was most severely tested during the Covid-19 pandemic, revealing how important it is gain public acceptance of novel technologies. One of the major factors affecting the public acceptance of fusion energy lies with the veracity of the science and motivations of those promoting it.

For fusion energy to be accepted by the public concerns such as those identified by its sceptics need to be addressed. Opponents of fusion energy, such as Greenpeace, argue that it has four major problems: it is dangerous, expensive, uncertain, and unnecessary. Public support for the pursuit of fusion energy requires each to be addressed.

4.1 Danger

Fusion energy is inevitably associated with the misfortunes of fission-based nuclear energy: disasters including Three Mile Island, Chernobyl, and Fukushima; the challenges of managing radioactive waste with half-lives of thousands of years; and with atomic bombs.

There are two potential risks with using nuclear fission reactors. The first is a risk of a runaway reaction causing a nuclear explosion. Nuclear fission reactors use enriched uranium 235, an isotope of uranium, as the fissionable part of their fuel. Nuclear fission reactors are fuelled every 12–24 months, which means highly radioactive fuel rods are loaded and then used to depletion over this period. Control protocols are used to keep the reactor working without it having a runaway reaction and going critical: their concern lies with the containment of the energy in the vessel. The propagation of an uncontrollable chain reaction can occur if a fission power plant loses its power supply to control the process, or if it becomes damaged (such as from earthquake, tsunami, faulty operation procedure, or deliberate attack).

The second risk relates to the management of nuclear waste. Spent uranium 235 is highly radioactive with a very long half-life (time to decay). It must be stored safely for several hundred thousand years, and it creates long-term nuclear hazards to manage. Furthermore, plutonium is one of the isotopes in the spent waste, and it can be used in production of nuclear weapons.

Should they be successfully developed, fusion power plants will be very different for three reasons. First, the most common fuel is likely to be a combination of deuterium and tritium, both are isotopes of hydrogen. Deuterium occurs naturally in water. Tritium is not found in nature and has to be produced (we examine how in Chap. 6). Tritium is used in nuclear medicine and biomedical research in areas such as heart disease, cancer, and AIDS, and defence technologies. It was painted on the hands on many brands of watches to make them luminous until the 1990s, when it was replaced by light absorbing materials (although some watch makers still use tritium in gas filled tubes). It has a short half-life of 12 years, making it much more feasible to manage and its emitted radiation is not nearly as damaging as uranium 235. Nonetheless, the neutrons released will leave radioactive materials in the surfaces of the containing vessels, and it will require infrastructure for handling and containment.

Second, tokamak vessels will only contain a few second's worth of fuel.[3] This means they do not contain enough fuel for a runaway reaction. Moreover, unlike fission which is virtually self-sustaining, the fusion process is difficult to keep going, and control protocols of fusion power plants will be required to sustain a fusion process: their concern will be with the release of energy in the vessel. The reaction of the basic chemical elements used for fusion energy can, unlike fission, be extinguished almost instantaneously.

Third, fusion will not create difficult to manage nuclear waste. Its fuel sources do not include the actinides—the radioactive metallic chemical elements—found in plutonium and uranium. As we shall see in Chap. 7, the STEP power plant in the UK has been given permission to be built on a non-nuclear licensed site and has a separate Fusion Regulator, distinct from the UK's Office for Nuclear Regulation that oversees fission reactors. Nonetheless, if fusion energy production becomes industrial scale, the management of radioactive material, particularly activated steel, will become significant and require substantial capabilities in remote robotic handling and maintenance.

So, although radioactive waste from fusion is very small compared to the fission processes using uranium and plutonium, and have limited military use in weapons, they are still associated by opponents with all the dangers of nuclear energy. Through the association with fission, there is also the concern about uncontrolled reactions occurring in vessels, although this is something that cannot occur with fusion: uncontrolled reactions and catastrophic meltdowns cannot occur.

The only way of addressing the misconceptions about fusion energy and its association with fission energy is for honest and continuous communication with the public, especially if it moves from the research sphere to industrial-scale production.

There are inherent dangers in all energy sources. Energy production has a high price in the air pollution and greenhouse gases it creates, and in the accidents caused as it is produced and distributed. It is estimated, for example, that five million people die prematurely annually as a result of air pollution.[4] In a review of the safest form of energy, Ritchie (2020) summarizes her findings: "by all metrics, fossil fuels are the dirtiest and most dangerous, while nuclear and modern renewable energy sources are vastly safer and cleaner". Coal and oil result in death rates thousands of time greater than nuclear and renewables. She estimates that based on the demand for one terawatt-hour of energy per year (roughly the demand of a town in the European Union of 27,000 people), if this energy were to be derived from single sources, it would cause the following number of premature deaths each year: coal: 25; oil, 18; and gas, 3. For that energy to be produced by wind it would take 29 years

---

[3] Ian Chapman, Kavli Lecture, Royal Society, 27th January, 2021.

[4] https://ourworldindata.org/air-pollution#air-pollution-is-one-of-the-world-s-leading-risk-factors-for-death.

before someone died; hydro-energy and solar, 42 years; and solar 53 years. Using nuclear energy, it would take between 14 and 100 years for someone to die.

Ritchie also provides data on the accident rates of non-carbon producing energies, and the risks of death in energy production is higher in renewables than in nuclear. That is death rates in fission reactors, not in fusion.

The conflation of fission and fusion also leads to concerns about military applications. Despite the significant differences between nuclear fission and fusion, many of those in the fusion energy industry are, revealingly, wary of the term "nuclear": the term fusion is rarely accompanied with a preceding "nuclear". While fusion energy is hitched unfairly to what is considered a difficult historical legacy in nuclear science, this unfairness is less immediately apparent when it comes to nuclear bombs.

As part of the Manhattan Project to develop nuclear weapons for use in World War 2, Robert Oppenheimer, the project leader, and Edward Teller, one of its leading scientists, recognized in July 1942 that when it cames to producing the most powerful bomb, fusion provided greater opportunity than fission. The technical difficulties of utilizing fusion led to the prioritization of fission, and it was fission that caused the explosive devastation of Hiroshima and Nagasaki. But work continued on fusion bombs, also known as thermonuclear or hydrogen bombs, during and after the War, with Teller reporting weekly to Oppenheimer on progress. Little headway was achieved, but following instructions from President Truman concerned about Soviet progress with a fission bomb in 1949, research efforts increased. By using X-rays produced by fission to detonate a fusion reaction, the first thermonuclear bomb was tested in 1952. Its explosion was equivalent to ten million tons of TNT, 700 times more powerful than the bomb dropped on Hiroshima. The Soviet Union tested its first thermonuclear weapon in 1955, with significant contributions from Andrei Sakharov, who spent the rest of his life warning of their dangers. The US Department of Energy's history of the Manhattan Project, upon which this account is based, concludes that as a result of the development of fusion weapons:

> There was now almost no limit on the size of an explosion either superpower could create. Fission bombs could only be built up to a certain maximum size, because only so much fissile material could be assembled without reaching a critical mass. Fusion bombs had no such limitation.[5]

The public's perception of fusion is sometimes associated with thermonuclear bombs. Bad memories and associations tend to linger and are hard to shift. The link between development of fusion is occasionally connected with the maintenance of nuclear weapons, for example, in the research at the National Ignition Facility (NIF) at the USA's LLNL. This is a unique experimental facility that addresses both military use for national security and energy research. It conducts work on safety, security, and reliability of the US nuclear stockpile through its Stockpile Stewardship Programme, avoiding the need to use underground nuclear testing. LLNL is also

---

[5] https://www.osti.gov/opennet/manhattan-project-history/Science/BombDesign/hydrogen-bomb.html.

home El Capitan, the world's most powerful supercomputer, which began operating at the end of 2024. It is capable of two exaflops of performance, or two quintillion (a billion billion) calculations per second. This computational power and speed is needed in predictive modelling to assess the safety and reliability of the US nuclear stockpile. It can also be deployed to model the almost instantaneous reactions occurring in inertial fusion.

Although it has not had responsibility for weapons since the 1950s, the UKAEA had nuclear weapons research in its original remit. The echoes of such association from the past can stay in public memory. That memory stays refreshed when mention is made of military applications when justifying expenditures on fusion research, such as that happens at NIF, and is argued to be the case with the Chinese fusion programme.[6] The view expressed by sceptics such as Makhijani (2024) who argues: "fusion research for peaceful use and military use are highly intertwined" needs to be addressed. Continual emphasis is also needed on the differences between fission and fusion, and tokamaks and inertial confinement machines.

It remains the case, however, that the development of hydrogen bombs led to a significant increase in scientific knowledge of fusion reactions, which although opened up the possibility of fusion energy, also ultimately enabled the development of a weapon of mass destruction.

4.2 Expense and Price

The development of fusion energy is undoubtedly expensive, and opponents point to the amount of immediate energy that could be produced by investing in renewables such as solar and wind, rather than fusion. They consider the comparative economic costs of renewables would be significantly discounted if they were to receive the same level of government investment as has fusion research. This begs the question of whether what economists call the externalities—the complementary benefits—of investment in renewables is comparable to those in fusion. Investment in fusion has been seen to have benefits for industrial development as well as progress in science and technology in related areas, such as robotics and new materials, some of which will improve the technical performance and economics of renewables and beyond. These benefits need to be included in assessing the cost of developing both renewables and fusion energy.

A large part of the costs of fission reactors lies with dealing with radioactive waste and decommissioning old reactors. Fusion reactors produce small amounts of radioactive waste which are relatively easier to store than waste from fission reactors, and decommissioning old fusion reactors would be significantly less

[6] "Ferreting out the truth about fusion: Interview with Bob Rosner". *Bulletin of the Atomic Scientists*, 80(6), pp. 345–351. In this interview, Rosner argues that one advantage of inertial confinement experiments for military purposes is that it prevents the need for the underground testing of nuclear weapons.

expensive. Lessons from the decommissioning of JET will also feed into designs of machines to optimize their life time cost.

A study of the cost competitiveness and commercial viability of tokamaks suggests that at current modelled projections based on present fusion energy projects, they will not be competitive: calculating that they will produce at $150/MWh at 2020 price, whereas to be competitive they will need to produce at $80–100/MWh. However, it contends that trends in design, such as reduced size of plants, and economies of scale in production, have the potential to make fusion energy commercially competitive (Lindley et al. 2023).

In the end, it comes down to price, and fusion energy will need to compete with renewables, such as solar and wind. As Berkhout puts it:

> Grids are also being transformed to integrate renewables, becoming more two-way, adaptive and networked, with trade over longer distances. There remain problems with grid infrastructure investment, but this will come. Price will be decisive and the accelerating competitiveness of renewables, including grid network effects—we are still in the early stages of this industrial revolution—mean that follower technologies, like fusion, will need to demonstrate quite clear advantages, not just in availability but also in price.[7]

4.3 Uncertainty

The viability of fusion energy is uncertain, especially given its lengthy time horizons. Opponents argue that by the time fusion energy comes on stream, if it does, the climate change crisis will be too advanced to be addressed. Proponents of fusion contend there are often extended periods of many decades between fundamental scientific breakthroughs and any practical applications, and given the complexity of the fusion project, its timescales are not unexpected.

That said, performance in fusion energy is improving markedly. A catalogue of fusion projects around the world reveals that progress—measured by the duration of plasma confinement and power output—actually continues to exceed the rate of Moore's Law, which predicted the doubling of transistors on a chip every 2 years.[8] The fundamental challenges of fusion energy remain however: how to produce more energy than is put in to create it, and then to safely channel that energy at competitive prices. And doing that remains uncertain and leads to the view of one sceptic that:

> Climate change is a serious problem which needs to be addressed by "decarbonizing" our energy generation systems sooner rather than later—by 2050 at the absolute latest. And we're not going to have practical fusion energy in that kind of time frame. There's no way.[9]

---

[7] Berkhout, F. (2024), personal communication, 26th January.

[8] http://laurencehunt.blogspot.com/2018/07/is-compact-fusion-power-now-fewer-that.html.

[9] https://thebulletin.org/premium/2024-11/ferreting-out-the-truth-about-fusion-interview-with-bob-rosner/.

The quest of moving from research machines to actual powerplants is currently being assisted by developments in design and manufacturing, including digital, data modelling and AI, high-temperature superconducting magnets, and new materials and heating systems. While these technologies help reduce the uncertainties around the development of fusion energy, they do not remove them.

In addition to the challenges of designing a commercial fusion machine itself, there remains a question of how in practice it feeds an electrical distribution network. The intent of fusion energy is to provide continuous base load, yet fusion machines deliver intermittent energy pulses, so there is a need for some form of energy storage to effectively buffer the conversion of energy pulses into a steady heat source for base load electricity generation. One option based on current technology is the use of molten salt heat storage.

Apart from the use of new design and manufacturing technologies, the question of whether there will be practical fusion energy by 2050 will depend, as we argue throughout this book, on the levels of resources dedicated to it, and the intensification and effectiveness of fusion innovation ecosystems that lead to its industrialization.

4.4 Necessity

The argument that fusion energy is unnecessary is primarily based on the assumption that by the time it contributes all necessary energy will be provided by renewables. This presumes that there will be sufficient capacity to store renewable energy to cater for baseload. Whether battery storage will be sufficiently advanced, and whether their materials and manufacturing processes will be appropriately environmentally friendly, is itself highly uncertain and remains an open question.

Blanket opposition without consideration of the profound differences between fission and fusion energy misjudges the environmental credentials of the latter. As well as the argument that fusion will provide a vital supplement to energy from wind, Sun, and tide, some consider fusion energy is itself a renewable energy source. Fusion energy will undoubtedly produce radioactive materials that will need to be dealt with and processed, which is why proponents often discuss it as non-carbon-producing rather than "green". As well as being relatively clean and safe, a feature of fusion energy that may prove popular with the public is it is contained, that is it is produced in relatively small plants. Renewables, as observers of large solar and wind farms know, can sometimes assume significant environmental presence, with the destruction of natural habitats. Fusion plants can also potentially be located close to demand in densely populated areas, reducing the losses involved in electricity transmission and storage. Should the fusion energy industry develop, concern about its necessity may be mitigated by the jobs it creates and the opportunities it provides for rewarding careers.

Nuclear energy is an understandably highly emotive issue for many people, deeply concerned about the environment. An Ipsos/MORI poll conducted around the world in 2011 after the Fukushima disaster, found 62 per cent opposed to nuclear energy as a means of electricity production.[10] Such is the opposition to nuclear energy in Germany that it will be abandoned by 2022, yet coal-fired energy stations, which are also to be banned, will operate until 2038. But concerns about military weapons and radioactive materials have to be calibrated alongside worries about climate change, which is very high in the public consciousness: people are highly aware of rising temperatures around the world and the growing preponderance of extreme weather events such as tropical storms. Changes in weather patterns are widely appreciated by the public as a problem to be confronted, and addressing them is seen as a pressing social goal.

Some opponents of fusion energy wish to see declining global production of electricity. It is certainly to be hoped that numerous improvements in energy efficiencies will come on stream capable of reducing some demands. The argument for fusion energy, however, is based on the assumption that the global demand for electricity will increase. Electricity demand globally is expected to increase by at least 70 per cent by 2040.[11] The benefits of electricity for the quality of lives is inestimable, and the thought of living without it is unimaginable. Lenin said that communism required two things: soviet organization and the electrification of the whole nation. At the death of Thomas Edison, who commercialized electricity production and distribution, President Hoover suggested the American people turn off their lights for a minute to remember the world as it was before his innovations. It would remind everyone, he said, of the dependence of the country on electric energy for the quality of their life and health.

Despite necessary and welcome advances in energy-saving technologies and products, more energy will be demanded. Not to cater to that demand is to deny to hundreds of millions of people currently without electricity a key contributor to their progress.

As part of the public debate about the benefits and fears of fusion energy, persuasive arguments about why it is needed have to avoid hype: overselling fusion energy invites scepticism about its necessity. Milne (2020) includes fusion as a case study in her book about hype in science. Although sympathetic to the idea of fusion research, she cautions about being too positive or negative about its future. While extolling the virtues of science, she warns about its occasional "lack of understandable information, an abundance of over-simplified generalisation and indecipherable ambiguity". These conditions, she contends that "make hype ripe for unhinged propagation". She uses a number of cases, including the Pons and Fleischmann cold fusion experiments, as examples of hype surrounding fusion, and suggests that these setbacks may help explain why fusion isn't part of the mainstream public narrative. A narrative which essentially says that multiple sources of future energy are needed, and fusion could potentially contribute to the decarbonized energy mix of an enduring net zero society.

[10] https://ourworldindata.org/grapher/public-opposition-to-nuclear-energy-production.

[11] World Energy Outlook, 2015, Paris, OECD.

While accepting its extraordinary ambition, Milne is critical of ITER and the risks involved in putting so many eggs in one basket. She is also concerned about intellectual property ownership in the project, which is owned by all contributing nations. This, in her view, diminishes incentives for commercialization and may explain the comparative lack of venture capital interest in fusion to date. It also opens the possibility of the disproportionate scale of state-owned Chinese investment allowing that country to dominate future fusion energy production (this issue will be considered further in Chap. 8).

In line with the views expressed by some in this chapter, she outlines three major concerns about the future development of fusion. First, its economics: whether fusion plants will ever be cheap, simple enough to build, and easy to plug into the global energy grid. Second, the radiation associated with neutrons bombarding the walls of vessels and disposal of residual nuclear waste. Third, the total energy needed to produce sufficient energy surplus. Adding the energy needs of magnets and cooling systems, she suggests that a ratio of 50 times more output than input will be needed. Despite these concerns, and conscious of the need to avoid exaggeration, Milne nonetheless contends that: "if we can find a way of getting fusion to work, at scale and within our current energy systems, it would truly be one of the most incredible feats ever accomplished by humankind".

Public acceptance of the virtues of fusion require unexaggerated, transparent reporting of its potential and challenges. This requires continual refutation of concerns about its safety, constant affirmation of its necessity, frankness about its cost, and openness about its risks and uncertainties. Time will tell whether fusion energy will be viable. While its dangers are limited, there are significant uncertainties about its costs and timing. There are always great uncertainties surrounding all major innovations, but there is no doubt, given the extent of the climate crisis, that all potential options to alleviate it are necessary. The risks it poses need to be measured against the risks of uncontrolled climate change. Given that innovations diffuse more effectively when their source is trusted and they accord with social values, the onus is on governments, scientists, and those in the fusion industry to openly and honestly make the case for fusion energy as part of the battle against the existential threat of climate change.

References

B. Lindley, T. Roulstone, G. Locatelli, M. Rooney, Can fusion energy be cost-competitive and commercially viable? An analysis of magnetically confined reactors. Energy Policy **177**, 113511 (2023)

A. Makhijani, The entanglement of fusion energy research and bombs. Bull. At. Sci. **80**(6), 352–357 (2024)

G. Milne, *Smoke and Mirrors: How Hype Obscures the Future and how to See Past it* (Robinson, London, 2020)

R. Ozaki, M. Dodgson, Consumption of innovation, in *The Oxford Handbook of Innovation Management*, ed. by M. Dodgson, D. Gann, N. Phillips, (Oxford University Press, Oxford, 2014), pp. 271–289

H. Ritchie, *"What Are the Safest Sources of Energy?"?* (Our World in Data, 2020) https://ourworldindata.org/safest-sources-of-energy

H.G. Wells, *The World Set Free* (Macmillan & Co., London, 1914)

Part III
Innovation

Chapter 5
Using an Innovation Lens in the Fusion Energy Story

The best stories not only describe what happened, but in the telling enlighten how and why they happened. The story of fusion energy is enriched when it moves beyond an explanation of what happened and is happening and includes the questions of how and why the story is as it is. It needs some intellectual scaffolding, and to provide this we turn to the analytical lens of innovation. Innovation is a single, but broad-ranging, concept that can capture many of the factors contributing to the pursuit of fusion energy. Innovation underlies much social and economic progress and improvement in the quantity and quality of life. It is defined as the successful application of new ideas that add value, and it varies widely in its amplitude, from the incremental to the transformative. Various concepts from the innovation literature will be used to illustrate and illuminate the story of fusion energy. These concepts help understand its past, present, and future and, where they cannot explain, they raise and explore important unresolved questions. Innovation is an outcome and a process, and we use it in the former sense of machines and latter sense to capture the dynamics of the pursuit of fusion energy and the building of fusion innovation ecosystems. As a process, something dynamic, innovation is itself a story. The various concepts from the study of innovation described below help illuminate the conditions and circumstances that influence fusion's progress.[1]

Innovation commonly results from the intersection of a need and a possibility. The need for fusion energy is clear and obvious. It could provide a crucial element in the mix of future energy sources while being environmentally friendly. We need a safe, affordable, clean, and unlimited source of energy for generations to come. Fusion holds the possibility of contributing to the provision of that energy.

Many innovations are incremental in nature, but some are so profound and transformative that they change the world. Transformative radical innovations can

[1] Extended records of our understanding of innovation can be found in Dodgson et al. (2008, 2014); and Dodgson and Gann (2018).

M. Dodgson, D. Gann, *The Search for Viable Fusion Energy*, SpringerBriefs in Applied Sciences and Technology, https://doi.org/10.1007/978-3-032-13466-0_5

originate from a variety of sources and motivations. Some are responses to demand for improved products and services. Thomas Edison, whose innovations in electricity generation and distribution resulted in massive social and economic progress, adhered to the principle of only inventing with a market in mind. Other radical innovations, derived from basic scientific research, can be motivated more by the inquisitiveness of researchers than from a concern for any practical applications. Researchers in organic chemistry in the nineteenth century and solid-state physics in the twentieth century had little idea of just how profound the impact would be of their research on its respective development of the chemical and computer industries. Similarly, early researchers into fusion, such as Eddington and Oliphant, could not imagine the construction of tokamaks directing the phenomena they were discovering into sources of commercial energy. As in many other fields, fusion illustrates the value of curiosity-driven basic research.

Some innovations are underpinned not simply by research with a practical use in mind, or motivated by fundamental curiosity, but by an amalgam of them both. Donald Stokes (1997) called this intersection "Pasteur's Quadrant" in recognition of Louis Pasteur whose applied goals in his research coexisted alongside scientific creativity and rigour. While much of the research into fusion conducted in government laboratories and universities is "pure" basic research (typified by the Danish physicist, Nils Bohr), and is concerned, for example, with plasmas, their research also has "pure" applied elements (typified by the American inventor, Thomas Edison), associated, for example, with issues such as vessel design and heat management. On the other hand, while much of the research in small fusion firms is highly applied, focused on rapid production of prototypes, they also contribute to knowledge and publish research. Fusion energy is, therefore, a radical innovation characterized by a clearly articulated need and demand and has been based on exploratory research driven by a quest for fundamental knowledge. It involves basic understanding of plasma physics with the practical aim of decarbonizing energy production. Pasteur would recognize the kind of innovation fusion represents, if not its nature. This distinction has consequences for government policies for science and industry which will be explored in Chap. 8.

Radical innovations can take many decades to gestate and to have their full effect. Progress from basic research through industrial application to social and economic benefits involves decades of effort. Research into solid-state physics in the 1940s began to be incorporated into semiconductors and integrated circuits in the 1960s, which were assembled into widely diffused computers in the 1980s, and it took many years thereafter to impact productivity. As basic research is so long term in orientation and uncertain in outcome, it is most commonly funded by government, with private sector investment entering as applications for the science and technology become proven. Basic research underpins the development of important contemporary technologies, such as high-temperature superconductivity, robotics, and material science. While the journey of fusion power to-date can be characterized as movement from basic research to applied technology, innovation theory also emphasizes the iterative nature of these connections, with scholars such as Nathan Rosenberg (1982) showing how the development of technology can stimulate and direct science. This has implications for those impatient with the speed of development of fusion, and for the agendas of scientists which may be directed not at "pure" curiosity-driven research, but at explaining and solving practical technological problems. It also has implications for the development of prototype fusion plants, proceeding in their designs before research has all the answers to the issues they need to address.

These features of fusion correspond to theories of the economics of innovation. The high costs and uncertainties of fusion energy, along with its long time horizons, are characterized as *incomplete* and *asymmetric information*. In such circumstances, the private sector finds the risks too high to attract investment, and government assumes a role in investing in knowledge that is a *public good*, available to benefit all. To date, government expenditure on fusion has dwarfed that in the private sector, recognizing public good and what the economics literature calls *positive externalities*. That is, benefits accrue beyond the immediate investment that are not captured by the private sector. In the case of fusion energy, for example, complementary technologies in materials, robotics, engineering, manufacturing, and information systems are developed with broader applications for other industries. The externalities of fusion also include the benefits of not having to deal with the environmental costs of carbon-producing technologies. Government also plays a role in addressing *systems failure*, including creating the necessary *economies of scale* which could not be achieved by the private sector alone, and in facilitating *networks* of connections between research institutions and industry that often only exist through state encouragement. These concepts explain why fusion has to-date been a government-led development.

Joseph Schumpeter (1934), the doyen of innovation economists, described an entrepreneurial model of innovation by arguing that new developments in technology attracts "swarms of new entrants", many of whom will fall by the wayside and a few will grow and become industry leaders.[2] Fusion has not attracted swarms of start-ups, unsurprisingly given its lengthy timescales and high levels of investment required (and "incomplete and asymmetric information"), but it has invited a number of business initiatives with impressive ambitions and diverse approaches in science, technology, and business models, and some of their stories are told in Chap. 7. Around 40 start-ups in the field have attracted around $9 billion in private investment, including from high-profile entrepreneurs and investors such as Bill Gates and Jeff Bezos. Questions arise about the veracity of this theory in fields such as fusion when the scale and breadth of resources and capabilities needed for the development of a new industry or technology exceed those within entrepreneurial start-ups. It does, however, point to situations that may arise if in future the fusion industry does develop, to some of the advantages of entrepreneurial firms in their ability to quickly seize opportunities and take risks, and to how there will be winners and losers in such circumstances.

[2] Schumpeter analysed two models of innovation: one occurred through entrepreneurial efforts in new firms, the other was seen in large firms' investments in R&D.

Innovation theory offers the concept of the emergence of a *dominant design* in technology. During its early stages of development, there is great uncertainty about which technology, and particular design configurations, has the greatest potential. In this period of "foment", there are competing designs before eventually one comes to dominate in the market. Although one design, the tokamak, is the preferred option, at least as experimental machines; at this stage, competing designs, such as stellarators and inertial confinement are being actively developed, and very different configurations of reactors, in principle, size, and energy, are being developed based on an increasingly diverse range of concepts and methodologies. These competing designs help shape general knowledge about how fusion might work as well as creating important technical capabilities for the development of different types of machines, for example, through the development of powerful magnets, sensors to control performance, and robotics used in maintenance.

Another well-established insight from the innovation literature is the *S-curve* of a lifecycle of a product, service, or technology, and its impact on the market. Stages in the lifecycle includes an initial nascent period of slow progress; a period of rapid growth and acceleration; then periods of maturity and decline. With this framework in mind, fusion energy can be thought of as being at the nascent period, awaiting the stimuli to move it into the period of rapid growth. The successful development of an innovation commonly requires access to *complementary assets*, that is, the adjacent supports necessary for an innovation to diffuse. Electric vehicles, for example, require the complementary assets of recharging stations; fusion machines will require a range of supportive technologies in, for example, energy storage.

Innovation, for Schumpeter, was the result of "new combinations". Fusion energy involves the combination of some of the most remarkable developments in science and technology. It is very notable when you spend time with people working on fusion that they talk in the most matter-of-fact way about the most astonishing and diverse achievements. Creating temperatures of 150 million degrees Celsius is deemed commonplace. Moreover, these extraordinarily hot temperatures are reached in vessels 2 metres away from extremely cold temperatures of nearly absolute zero, minus 273 degrees centigrade. Mention is made almost casually about cameras being used with exposures of three billionth of a second. These achievements require the combination of advances in scientific and engineering knowledge in a wide range of fields, from new materials to new digital design tools. An implication of this, discussed more below, is that, for progress to occur, knowledge previously bounded within disciplines and professions needs to be shared, melded, and "fused" together .

Using the lens of innovation management also embeds the notion of *risk*, not as something unacceptable but essential to progress, critical to understand and deal with. And it brings with it the value of balanced portfolios of risk, part of which are highly risky with the potential of high returns. This shifts some of the language used to describe the pursuit of fusion energy as a "gamble" towards a more appropriate descriptor as an "option": that is an investment that once more information is available in the future gives the preferential opportunity to make further investments in line with reduced risk.

Further insights from the study of innovation apply to the conduct of *major projects*. ITER is by any measure a major scientific research project, and STEP is a major prototype development project. Major projects have an appalling record on their delivery, often due to their large number of interdependent components and subsystems, and are commonly well over budget and significantly late (Flyvbjerg et al. 2003). Innovation has a role to play in improving their performance (Davies et al. 2017). This includes in their construction of often highly complex facilities involving significant advances in construction technologies and practices, such as digital twins and off-site prefabrication. Producing complex equipment, often with new and advanced materials, can involve substantial advances in machining, materials handling, and assembly. Major projects also offer opportunities for innovation in management practices in areas such as partnering/alliances, subcontracting and risk sharing.

Essentially, innovation is needed in major projects to deal with their endemic uncertainty by capturing and using learning from previous projects, and to take advantage of new opportunities. When the project is constructing a scientific experiment, such as ITER, or first-in-the-world prototype, such as STEP, these uncertainties are compounded with fundamental questions such as: how much will it cost?, how do we make it?, and will it work? Furthermore, while the progress of individual components and subsystems can often be anticipated in advance, it is difficult to predict how the components will interact when joined together as a system because of the interdependencies amongst them and uncertainty caused by their coordination. Given the lengthy duration of such projects, new opportunities may arise, for example, in new materials or design and manufacturing processes, and the capacity to take advantage of such innovations can have a significant impact on their delivery. Innovation theory connects with the field of economics known as evolutionary economics, where in complex and uncertain circumstances—that is not "complicated", but "complex" in the sense that outcomes are unknown—then emergent approaches are best, with a focus on learning as conditions become clearer. Learning has long been known to be a central element of innovation and the collaboration that supports it (Dodgson, 1993a, b) and is a fundamental principle of organizations such as ITER and JET.

What this innovation research implies in practice is that the organization responsible for integrating components into a system has to establish a variety of standardized routines for dealing with stable, predictable and known risks, while having the flexibility to adjust plans and modify routines when conditions change. They have to maintain the standardized operating routines which are all-important for costs and efficient delivery when conditions are well-understood, recurring and predictable, while at the same time creating the innovation, flexibility, and responsiveness to address unexpected problems and opportunities (see, for example, Davies et al. 2016, 2017). They have to offer, what has been known for some time as the "disciplined flexibility" (Sapolsky 1972) to ensure that original project objectives are met within a changing and uncertain environment.

In the end, much of the development and use of innovation in large projects such as building a fusion machine comes down to learning from the successes and failures of previous projects. These include, for example, the way Airbus manages its supply chains across national borders and the techniques NASA uses for managing its intellectual property. Paul Methven, CEO of UKIFS, says the project is drawing on lessons from complex mega projects in many sectors and how its complexity and uncertainty is "a great opportunity to think about how best to do things and not to be overly constrained by the norms we may have become used to".[3]

In the light of the earlier observation about the importance of learning, he says:

> If I was going to encapsulate everything that STEP is about, you could just say it's learning by doing. We're learning by doing the design, we're learning by doing the evolution of the company, by doing the project.

Studies have long discussed national, regional, and sectoral "innovation systems", and more recently "innovation ecosystems", and in any analysis of fusion, recognition has to made of all the parties that contribute to its development, and very importantly to their connections. Innovation ecosystems are commonly defined as interconnected organizations that create an appropriate value through innovation (Autio and Thomas 2014). The contributing organizations include large and small firms, researchers, and governments; and these can be complemented by actors from finance and venture capital and from organizations such as Non-Governmental Organizations, philanthropists, and industry associations. The various contributions made to innovation ecosystems is shown in Fig. 5.1. No one actor in the ecosystem possesses all the required capabilities and resources for its effective operation: collaboration is sine qua non, especially in crises and when the challenges are complex.

The organizational antecedents of innovation ecosystems can lie with large firms, in small, entrepreneurial start-ups, or around a research base, as in the case in the

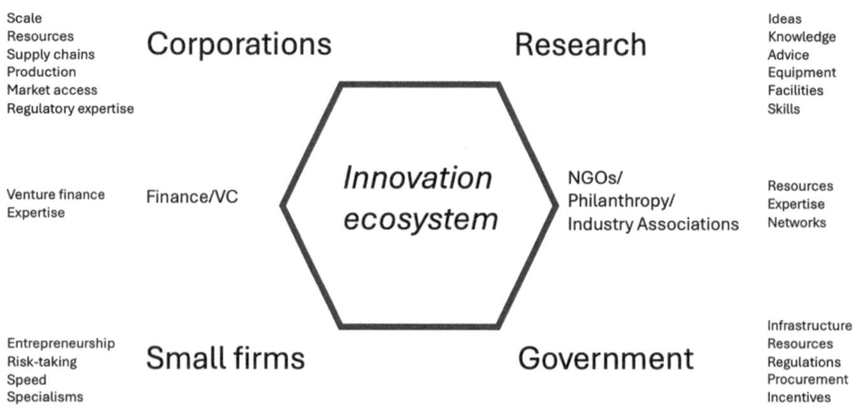

Fig. 5.1 Contributors to innovation ecosystems

[3] https://www.theengineer.co.uk/content/opinion/comment-the-challenges-and-opportunities-from-a-step-change-in-fusion-power/?utm_medium=Social&utm_source=LinkedIn#Echobox=1655195401-2.

UK around the UKAEA, and in the USA around laboratories such as Lawrence Livermore and universities such as Princeton and MIT.

Interactions between many actors are needed for fusion to deliver viable energy, including large research institutes and universities, small start-ups, governments, and regulators, suppliers of equipment, construction firms, investors, scientists, engineers, and designers. Successful innovation at the scale of fusion energy requires all these players, and importantly needs them to work collaboratively, cohesively, and in complementary ways. The history of systems innovations also shows that competition between various experiments and projects creates healthy tensions that attract further interest and attention, broadening knowledge about different design options. Understanding the path to fusion energy as the development of an innovation ecosystem provides an opportunity to appreciate its history, and the ebbs and flows of its research and engineering, economic and environmental considerations, and political and organizational issues. It focuses our attention on the multi-disciplinary, multi-dimensional phenomena that when they come together successfully in extraordinary experiments they eventually change the world.

Finally, the study of innovation shows the importance of *serendipity* and *luck*, *chance,* and *random events*, all of which, as we shall see, have played a role in the development of fusion energy. The innovation lens on fusion allows us to capture these crucial human features in its development as well as the scientific, technological, and economic factors. The risks and uncertainties in experiments and innovation in general commonly leads to failure, in not meeting objectives, budgets, or timelines. Just as Helios could be distracted by familiar comforts or distracted by nymphs on his journey bringing the Sun to earth, the innovation journey can be diverted by many appealing, but ultimately misleading, alternative directions. Failure, however, is often a means to learn, and progress is also hostage to chance events.

5.1 Failure

Charles Seife's (2008) history of fusion tells of a sequence of missteps if not outright disasters in the design and application of fusion machines. He reports on how the Princeton Plasma Physics Laboratory's National Compact Stellarator Experiment, tens of millions of dollars over budget, and years behind schedule, was closed down in 2008, and the Mirror Fusion Test Facility, at Lawrence Livermore National Laboratory, was defunded in 1986 without being used even once. The US Department of Energy reported in 2016 that despite the $3.5 billion invested in the NIF over many years it had to date failed to ignite a self-sustaining fusion reaction although this was achieved in 2022.

Zhong (2024) points to the way:

> Fusion's history is a graveyard of missed deadlines and thwarted milestones, bursts of excitement followed by bruising disappointments.

Seife does not hold his punches in explaining such misfortunes, referring to scientific arrogance, overselling to politicians, and in some cases outright delusions.

In science and innovation, however, "failures" are commonly precursors to success. Lessons from experiments that fail to deliver on expectations provide valuable information not only on why they don't work as expected, but what might subsequently succeed. Each failed experiment may be expensive, time-consuming, and organizationally, professionally, and personally embarrassing, but may lead to crucial advances that would not have occurred in their absence. Each experiment, success or failure, can be a step towards the emergence of a dominant design and potentially eventually contribute to a new industry.

Project ZETA is described by Seife as a British embarrassment and failure, and it sometimes goes with the sobriquet "the ZETA fiasco". ZETA (Zero Energy Thermonuclear Assembly) was an early experiment in fusion research. It went into operation in 1957 at the UKAEA's Harwell site and was at the time the world's largest and most powerful fusion machine. Despite changes to its design during its assembly, it was delivered on time and on budget and very quickly produced encouraging results, releasing millions of neutrons in five-million degree temperatures. Plasmas were created in Zeta that lasted milliseconds, whereas in previous machines had rarely lasted more than a few microseconds (a millisecond is a thousand microsecond), so this was an order of magnitude improvement.

This positive information began to leak out, and in January 1958 a press conference was held beside the ZETA machine at which 400 journalists attended. John Cockcroft, the Nobel Prize winner who headed British nuclear research and was responsible for the Zeta program, announced "Neutrons have been observed in about the numbers to be expected if thermonuclear reactions were proceeding".[4] Despite the element of caution he added, the message was interpreted by the press that fusion had been achieved (unsurprisingly as he said he was "90 per cent certain" it had happened). This outcome was reported directly to the Prime Minister, Harold Macmillan. Having recently suffered the political humiliation of the Suez Crisis, the British press was keen to celebrate how the nation remained a world leader in science. In banner headlines on their front pages, newspapers such as the Daily Mail heralded "THE MIGHTY ZETA", the Daily Herald trumpeted "Britain's H-men make a SUN", and the Daily Sketch boasted that Britain had created "A SUN OF OUR OWN". The journal *Nature* reported that experiments had led to thermonuclear fusion, and the Financial Times wrote a lengthy article on the project.

In short order, however, leading American and Russian nuclear scientists, Lyman Spitzer and Lev Artsimovitch, both questioned the ZETA results and, indeed, pointed out their theoretical impossibility. Four months after the press conference, the Harwell team issued a retraction, explaining the neutrons were produced from

---

[4] Hendry and Lawson (1993) report that one motive for such an early announcement was the pressure to demonstrate priority over American researchers. They also point to the context at the time of the success of the Russian launch of its Sputnik satellite (contrasting with the American's launch which exploded on the ground), and the fire and release of radioactive material at windscale the nuclear energy plant: there was pressure for a good news story about nuclear research.

instabilities in the plasma, rather than from fusion. The "embarrassment", "failure", and "fiasco" of this retraction probably related more to public relations than scientific achievements, but given the prominence of the publicity surrounding ZETA, it was a major setback for Harwell and for fusion research. The Daily Express summarized the feeling that "this is a great blow for British prestige".

ZETA researchers, however, felt they had produced a great pioneering experiment.[5] Indeed, important lessons were learned from ZETA, which continued operating until 1968. High temperatures were achieved, understanding of how plasmas could be constricted in toroidal vessels improved, and shortcoming in the design of the reactor were revealed, and subsequently overcome in later tokamaks. The project led to advances in the measurement of high temperatures using lasers, which latterly proved extremely important in validating the tokamak concept in the USSR. At this time, fusion research was conducted in secrecy, and scientific results were not available for peer review. This episode also led to the removal of the secrecy requirements around fusion research, allowing greater engagement with university research in the UK, and improved opportunity for international collaboration.[6]

There have been cases of deserved failure in fusion research, most obviously in the celebrated case of a German, Ronald Richter, working in Argentina. In 1951, based on Richter's claims, Argentina's President, Juan Perón, announced that the country had "completed thermonuclear reactions under conditions of control". Such fanfare did not survive for long as, within a year, Argentinian scientists showed such findings were completely faked. But most "failures" have advanced the science and technology of fusion energy.

5.2 Chance

During the late 1950s, the concept of a common large-scale project began to be discussed by European fusion physicists. The Council of the European Community decided to support this ambition, and the Joint European Torus (JET) was announced in 1971. The initial design of the project, undertaken between 1973 and 1975, required the team to draw all the necessary plans for launching the project, perform preliminary testing, and to start building some crucial prototypes, such as a toroidal coil and a vacuum vessel sector. At the invitation of its Director, Bas Pease, the design team, led by Paul-Henri Rebut, was based at the UKAEA's Culham site.[7]

[5] Roland Pease, "The Story of 'Britain's Sputnik'", BBC Radio Science Unit, 15th January, 2008.

[6] One of the reasons for the introduction of secrecy was in 1950 Harwell's Head of Theoretical Physics, Klaus Fuchs, admitted to being a spy who had fed American and British nuclear secrets to the Soviet Union. Fuchs was fully aware of the state of fusion research.

[7] Culham had already been selected as a site for ICSE (Intermediate Current Stability Experiment), a controversial and subsequently aborted successor to ZETA, that was entirely British funded.

The decision of where to locate JET was contentious.[8] Five sites were proposed: Culham, Cadarache in France, Garching in Germany, Mol in Belgium, and the European Commission proposed Ispra in Italy. The five were eventually reduced to two: the UK and Germany. The UK, which had joined the European Community in 1973, was keen to host the facility and was supported by France, Denmark, Ireland, Netherlands. Germany and Luxembourg favoured Garching. Belgium and Italy abstained. The disagreement over its location threatened the whole initiative.

On the 13th October 1977, Lufthansa Flight 181 *en route* to Frankfurt was hijacked by members of the Popular Front for the Liberation of Palestine seeking release of imprisoned Red Army Faction leaders, and flown after a harrowing number of stops to Mogadishu, Somalia. The pilot of Flight 181, Jürgen Schumann, was executed by the hijackers. On 18th October, the aeroplane was stormed by the West German counter-terrorism group, and all 90 hostages were released. Three of the four hijackers were killed. The German rescue team was supported by two members of the British Special Air Service (SAS); its leader had been trained by these elite special forces, and it is reported that the German team was supplied by a special type of grenade used by the SAS.

On the same day as the rescue mission, news of its success was relayed to a meeting of the British Prime Minister, James Callaghan, and German Chancellor, Helmut Schmidt. The agenda of this pre-arranged meeting included the Middle East and the location of the European Torus. Reports are of a friendly and respectful meeting between the two leaders. Schmidt expressed his gratitude for British support in the rescue; Callaghan replied that Germany was fighting terrorists threating worldwide democratic freedoms. Callaghan emphasized his government was eager to host a joint European device to demonstrate its strong commitment to the common idea of Europe. They both agreed that the decision on JET should not be divisive and should result from consensus.[9]

The decision to locate JET at the UKAEA Culham site was made 2 days after the Anglo-German summit. It may be too stark to directly attribute the decision to British assistance with the rescued hostages, but there is no doubt it influenced the "atmospherics" of the meeting, reducing contention and enhancing the goodwill in search for a solution for the location of the facility. Bas Pease's obituary refers to the way that the Germans conceded JET should be constructed at Culham in a moment of gratitude.[10] In his history of JET, its Head of Design, Deputy Director, and eventually Director, Paul-Henri Rebut, similarly refers to gratitude being a factor (Rebut 2017).

JET became the most important fusion research experiment in the world. That its history is linked to a murderous terrorist hijacking is an extreme example of the

---

[8] There was discussion during the late 1950s about locating future fusion research at CERN, something the British opposed (reported in Hendry and Lawson 1993). The complex decision-making process surrounding the location and configuration of JET are explained in Rebut (2017).

[9] https://www.euro-fusion.org/news/hijacking-a-jet/.

[10] www.independent.co.uk/news/obituaries/bas-pease-29570.html.

chance events that can have such influence on the development of economies and technologies (Beinhocker 2007).

As a further example of how good fortune matters to progress, government approval for the STEP programme to commence procuring its partners, following a long and detailed decision-making process, was granted just a few hours before a national election in the UK was announced. A delay of a few hours would have meant a deferral in decision-making during the election period, and no guarantee of approval under a new, incoming government with potentially different priorities and decision-making processes.

References

E. Autio, L. Thomas, Innovation ecosystems: implications for innovation management, in *The Oxford Handbook of Innovation Management*, ed. by M. Dodgson, D.M. Gann, N. Phillips, (Oxford University Press, Oxford, 2014), pp. 204–228

E. Beinhocker, *The Origin of Wealth: Evolution, Complexity, And the Radical Remaking of Economics* (Harvard Business Review Press, Cambridge, MA, 2007)

A. Davies, M. Dodgson, D. Gann, Dynamic capabilities for complex projects: The case of London Heathrow terminal 5. Proj. Manag. J. **47**, 1 (2016). https://doi.org/10.1002/pmj.21574

A. Davies, M. Dodgson, D. Gann, S. MacAulay, Bringing innovation into large, high risk projects. MIT Sloan Manage. Rev. **59**(1), 73–78 (2017)

M. Dodgson, Learning, trust and technological collaboration. Hum. Relat. **46**(1), 77–95 (1993a)

M. Dodgson, Organizational learning: A review of some literatures. Organ. Stud. **14**(3), 375–394 (1993b)

Dodgson, M. and Gann, D., Innovation: A Very Short Introduction, 2018. Oxford, Oxford University Press, 2

M. Dodgson, D. Gann, A. Salter, *The Management of Technological Innovation: Strategy and Practice* (Oxford University Press, Oxford, 2008)

M. Dodgson, D. Gann, N. Phillips, *The Oxford Handbook of Innovation Management* (Oxford University Press, Oxford, 2014)

B. Flyvbjerg, N. Bruzelius, W. Rothengatter, *Megaprojects and Risk: An Anatomy of Ambition* (Cambridge University Press, Cambridge, 2003)

J. Hendry, J. Lawson, *Fusion Research in the UK 1945–1960* (Harwell, AEA Technology, 1993)

P.-H. Rebut, The joint European torus. Eur. Phys. J. H **43**, 459–497 (2017)

N. Rosenberg, *Inside the Black Box: Technology and Economics* (Cambridge University Press, Cambridge, 1982)

H. Sapolsky, *The Polaris System Development: Bureaucratic and Programmatic Success in Government* (Harvard University Press, Cambridge MA., 1972)

J. Schumpeter, *The Theory of Economic Development: An Inquiry Into Profits, Capital, Credit, Interest and the Business Cycle* (Harvard University Press, Cambridge, MA, 1934)

C. Seife, *Sun in a Bottle: The Strange History of Fusion and the Science of Wishful Thinking* (Penguin, New York, 2008)

D. Stokes, *Pasteur's Quadrant: Basic Science and Technological Innovation* (Brookings Institute, Washington, 1997)

R. Zhong, *The Quest of Build a Star on Earth* (New York Times, 2024)

Chapter 6
Key Players in Fusion Innovation Ecosystems: Research Institutions

The first mover in all fusion ecosystems has been research. This chapter examines three of the larger fusion research institutions in leading fusion research, ITER, the Princeton Plasma Physics Laboratory in the USA, and the UKAEA. The first case illustrates the political difficulties surrounding scientific research at massive scale and poses the question of the dangers of the "sunk costs fallacy". The second case describes the challenge of attracting secure funding, and the third analyses the development of cumulative knowledge over many years despite the restructuring of the institutional arrangements surrounding fusion research. Judging progress in fusion has to account for these important and challenging contextual factors which affect their contributions to evolving fusion energy innovation ecosystems.

The following chapter considers the role of government and industry, with especial focus on the contribution of small fusion energy firms. It concludes with more consideration of the STEP programme as a catalyst for an innovation ecosystem. We start with the contributions of research institutions.

6.1 ITER

6.1.1 Background

Two years after referring to the Soviet Union as the "evil empire", and a year after he joked about outlawing and bombing Russia, President Ronald Reagan met his Russian counterpart Mikhail Gorbachev in Geneva. They mainly discussed the accelerating dangers of the nuclear arms race, but the final item in their communique of the meeting stated:

> The two leaders emphasized the potential importance of the work aimed at utilizing controlled thermonuclear fusion for peaceful purposes and, in this connection, advocated the

M. Dodgson, D. Gann, *The Search for Viable Fusion Energy*, SpringerBriefs in Applied Sciences and Technology, https://doi.org/10.1007/978-3-032-13466-0_6

> widest practicable development of international cooperation in obtaining this source of energy, which is essentially inexhaustible, for the benefit for all mankind.[1]

The peaceful use of nuclear power had already been on the political agenda for decades. In his address to the United Nations General Assembly in 1953, US President Dwight Eisenhower talked about the dangers of the proliferation of nuclear weapons and proposed the creation of an international atomic energy commission under the aegis of the UN. He proposed that:

> Experts would be mobilized to apply atomic energy to the needs of agriculture, medicine and other peaceful activities. A special purpose would be to provide abundant electrical energy in the power-starved areas of the world.[2]

Political interest in fusion at the highest level is also seen how in 1956 a Soviet delegation to the UKAEA's Harwell facility was led by the Soviet President, Nikita Khrushchev.

Scientists and policymakers had also been active for many years. The first international conference on the peaceful use of nuclear energy was held in Geneva in 1955, and 25,000 attended. In 1957, the European Atomic Energy Community (EURATOM) was established to develop and distribute nuclear energy in Europe.[3]

While there were many factors contributing to increased international attention to fusion, including growing scientific consensus, it was the Reagan-Gorbachev meeting that stimulated and catalysed action at scale. It led to the creation of ITER.

From the start, ITER, a scientific project, has had a powerful political overlay. It was recognized that a project on this scale—with an original budget of $10 billion—was beyond the means of individual nations, and collaboration was necessary. ITER originally had four members—Soviet Union, the USA, Europe, and Japan—and currently has seven members (known as Domestic Agencies): the 27 countries of the European Union, Japan, South Korea, Russia, the USA, China, and India, 35 countries in all. The UK's position was complicated after it left the European Union in 2020, and it remained engaged with ITER through an agreement called Fusion for Energy, part of Euratom, until the decision was made in 2023 to not renew its membership but to seek to continue to work with ITER.[4]

France, Japan, Canada, and Spain offered to host the facility, and competition for its location came down to a site in France, attracting the objection that it was too far from a port for delivery of components, and Japan, criticized for being an

---

[1] Joint Soviet-United States Statement on the Summit Meeting in Geneva, November 21, 1985, https://www.reaganlibrary.gov/research/speeches/112185a.

[2] Address by Dwight D. Eisenhower, President of the United States of America, to the 470th Plenary Meeting of the United Nations General Assembly, *eighth December 1953.*

[3] It is organizationally separate from the European Union, but consists of the same members plus Switzerland and is governed by some EU institutions. The UK left in 2020.

[4] See, for example, https://www.newscientist.com/article/2419671-uk-spurns-european-invitation-to-join-iter-nuclear-fusion-project/. The relationship with ITER has undoubdedly been muddied by the political uncertainties and confusions around the extended Brexit negotiations, and which have generally been argued to have had a negative effect on UK science.

earthquake zone. Being the host would attract many benefits, and avoid accusation of domestic taxpayer's money being used to support science and industry overseas. The USA backed Japan as a site, but the eventual winner was Cadarache, near Marseille in France. As compensation, Japan received funding for associated research facilities, guarantees of manufacturing projects, and assurances that ITER's first Director would be Japanese. Spain and Canada also received benefits in the form of projects and offices (Canada left the project in 2003). As a result of its location, around half ITER's budget (45 percent) comes from Europe, with the other parties contributing around 9 percent each.

The ITER project has had to proceed in the face of extraordinary political shifts and broader tensions between its members, including following Russia's invasion of Ukraine. The 1990s saw the collapse of the Soviet Union, lengthy economic recession in Japan, declining oil prices (making the search for alternatives less attractive), and a decision by the USA to withdraw from the agreement in 1996 (re-joining in 2003). It has proceeded amongst tensions over trade, including over intellectual property rights, when its principle is that all intellectual property obtained is available equally to all seven members. Commitment to fusion research has waxed and waned within the Domestic Agencies, with some experiencing substantial cuts to budgets in science and energy. Opposition has come from within the scientific community, especially from those whose own fusion research lost funding to ITER, and those not wedded to the idea of tokamaks being the best way forward. Those opposed to ITER include some green groups, trade unions, and two French Nobel Prize-winning physicists.

6.1.2 Scale

The scale of ITER can perhaps best be calibrated by the size of the cryostats being used around the vessel: they are 10 storeys high. To reach the site from the nearest seaport, a special network of 104 km of road has been established, with reinforced roads and bridges, and redesigned roundabouts. Heavy load transporters carrying over 1000 tonnes are being used, travelling at 5 km an hour.

Around 3000 people work at ITER, with around 15,000 contributing worldwide. It is supported by 15 other facilities around the world, testing and developing related science and technology, and is part of a related, larger international fusion research community. This includes EUROfusion, for example, which is a consortium of fusion research centres in the EU, plus Switzerland and Ukraine, and organizes and funds European fusion research in over 100 laboratories and supports nearly 2000 researchers.

Major components are being manufactured in Italy, Spain, Germany, Japan, South Korea, and India. The massive magnet coils, for example, are being built in Europe (14); Japan (9); Russia (1) and China (1). The central solenoid is being manufactured in the USA, and the cryostat is being made in India. Despite major efforts to ensure quality control and integration of components—all suppliers, for example, use the same stainless steel from a company in France—international coordination has proven extraordinarily challenging. As Khatchadourian (2014) put it, the project's success may rest on a simple question: Will everything fit together? And questions whether the design and the manufacture of the most sophisticated parts have been split apart in ways that are politically expedient but might be at odds with engineering prudence. From the outset, he says, each Domestic Agency vied to build the machine's state-of-the-art components, so that its industries could gain the know-how. The project faces immense challenges of systems integration, the greatest difficulty facing large, complex projects (Davies et al. 2017).

ITER's construction projects have been so large that it has invited many international consortia to share risks and deliver them. This has entailed the development of long-term, cross-cultural partnerships, an especially difficult form of collaboration (Dodgson 2018; Bardin 2020).

6.1.3 Major Challenges and Criticisms

ITER has been criticized on many fronts; for being too expensive, too slow, too complex, and being environmentally damaging.

On the question of cost, while ITER is extremely expensive, it is very hard to calculate what its final budget will be. This is partly because 90 percent of the contribution by members is "in-kind", that is provision of goods and services rather than money. Exchange rate fluctuations add to the difficulties of assessing contributions over long periods. By 2023, the project is estimated to have cost $22 billion.[5] Costs have to be defrayed against economic benefits. There are significant local economic benefits, and construction and engineering contracts have benefitted international firms. Claessens (2020:154) argues that:

> Companies state that working for ITER has helped develop new state-of-the-art technologies, improve their production and other processes, access business opportunities outside of fusion, and create synergies and new opportunities.

Claessens (2020) reports that the two French Nobel Prize-winning physicists have criticized ITER on cost grounds. Pierre-Gilles de Gennes says the project is too costly and underestimates the radiation problems, and George Charpak called ITER a "useless and overpriced reactor", and he estimates the cost of ITER will be somewhere between the huge variation of 13 and 41 billion Euro. If it is at the higher level, this is approximately quadruple its initial estimate. One possible way of calibrating this cost is to compare it with other big science projects. The Large Hadron Collider at CERN is estimated to have cost around 5–6 billion Euro, while the cost of running the International Space Station for 10 years is around 100 billion Euro. Making such comparisons, however, needs judgement calls on whether the

[5] https://www.scientificamerican.com/article/worlds-largest-fusion-project-is-in-big-trouble-new-documents-reveal/.

right decisions were made in designing and building these facilities and the extent of the potential benefits to be derived. As from its beginning, decisions are made on political as well as economic and scientific bases. It is also the case that ITER is being built under full French nuclear reactor regulations, adding significantly to cost.

On the positive side, the cost of ITER is spread across 35 nations (France making the largest contribution to the European budget). The study of innovation management shows the value of a portfolio of investments along a number of business horizons that defend and sustain existing business, diversify into new, related areas, and explore completely new departures. Balanced portfolios have a proportion of their investment in highly speculative, potentially transformative projects. Taking the highest estimate of ITER's *total* cost, this is 1.5% of the *annual* global investment in R&D.[6] It is possible to argue, given its potential, that this might appear to be a sensible investment if such a mythical thing existed as a balanced global scientific research portfolio.

On the question of delays, all large projects are characterized by failures to meet schedules (Flyvbjerg et al. 2003), and there is a history of delays in delivering fusion scientific equipment. The UK's Harwell research institute complained in 1955 about its major industry partner's severe overruns in costs and schedule in providing particle accelerator equipment. ITER's tokamak commissioning was first planned for 2016. This was delayed in 2010 to 2019. In 2015, this was further extended to 2025, with full experimentation being planned to be undertaken sometime between 2035 and 2040. Numerous factors have contributed to these delays, including external issues beyond ITER's control. So, for example, a tsunami in Japan delayed its contribution, and hence the whole project, by over a year. And a collapsed bridge in Italy disrupted supply from a manufacturing plant. Delays in payment—US and Indian cash contributions are, for example, reputed to be 2 years behind—add to the difficulties in planning. And there are many factors internal to the project that have caused delays. Late design changes to equipment and buildings delayed construction. Quality problems have arisen in construction and production of components. Contracts have been signed late, and ITER began with no templates for procurement or ways of ensuring all international contractors used the same procedures and quality assurances; and these needed to be developed.

Problems emerge simply because of the complexity of the project, which "pushes science and technology to its limits" (Claessen, 2020). That being said, the ambitions for ITER have scaled back from its initial intent. Originally the expectation was to achieve ignition and sustained burn in the reactor: that is generate energy by exciting the plasma to a state where it will heat self-sustainably. Presently, ITER is building a research demonstration facility, rather than a power plant. Any heat produced will be cooled then released into the atmosphere. A result of budget constraints and concerns about the science itself, Charles Seife describes this as a move to "ITER-lite": "a massive, more expensive version of what everyone else had

[6] Assuming global global R&D investment of $3.1 trillion.

already built". He also contends, however, that it is "the most ambitious international scientific project ever attempted" (Siefe 2008). Should ITER prove successful, there are proposals to create a fusion power station known as DEMO (Demonstration Power Station), followed by a commercial station, creating electricity for the grid, known as PROTO.

There are numerous scientific and technological unknowns in the project, including for example the continuing issues of the reaction of the materials in the vessel to radiation and the replacement and disposal of the tokamak's walls. For the reaction in the vessel, the facility will require 10 percent of the world's tritium, and the—as yet unachieved—ability to manufacture tritium at scale will be essential.

There have been continual criticisms of ITER's leadership and management, including basic questions about the appropriateness of scientists leading a 20-plus-year construction project. Khatchadourian (2014) reported that many engineers and physicists at ITER believed that delays were self-inflicted, having little to do with engineering or physics and everything to do with the way that it is organized and managed. He describes how a building was cancelled to save money, but alternative arrangements proved too expensive and it subsequently proceeded at greater cost. Claessen (2020) cites a senior manager who admitted that problems faced were more to do with management and logistics than they were technical.

Such criticism has a long history. When one of the early Directors of ITER's design, Paul-Henri Rebut, was removed from his post in 1994 (upon American insistence), he reported to Congress that the ITER organization had insufficient legal authority and independent funding, and a leadership of incompetent bureaucrats. He argued that the focus on consensus made decisions by the parties based on the lowest common denominator, and national representatives were more concerned with gaining returns to their contributions than the overall design of the project (Khatchadourian 2014).

A comprehensive review of the ITER Organization management was undertaken in 2013.[7] It offered some praise, referring to broad support and commitment to the project's success, capable and motivated staff in a number of technical and administrative areas, a number of very good best practices in a limited number of areas, and positive interactions and collaborations amongst Domestic Agencies. However, it said the project was in a malaise, with a lengthy list of criticisms, including how the inadequacy of the initial design and distributed manufacture of common components made it very difficult or impossible to make progress at an optimal pace; the lack of strong project management culture; insufficient numbers of people with large project and industrial experience; and ill-defined and poorly implemented decision-making processes. Adding to this list of woes, the report concluded that the organization had not operated as an efficient or effective management organization, with poor communications, limited incentives to reduce costs or risks, siloed management, and excessive bureaucracy.

[7] Madia & Associates, 2013 ITER Management Assessment, Final Report, October 18th, La Quinta, California.

The review made 11 recommendations, including: creating a project culture; finding a new Director; reducing the number of senior managers; strengthening systems engineering; instilling a nuclear safety culture; developing a realistic project schedule; better aligning interests with the Domestic Agencies—simplifying and reducing bureaucracy—and using Human Resource Management as a strategic asset. The US Congress made it a condition of its continuing funding that all these recommendations were implemented.

All these reviews and lessons provide valuable insight into the potential pitfalls for current and future fusion projects, whether experiments, prototypes, or energy plants. Just as knowledge of science and technology progresses, project to project, so too does knowledge of management and organization. Decades of research into innovation reveal how its major challenges often lie not with science and technology, but with management and organization.

6.1.3.1 Assessing the Value of ITER

ITER has significant political, scientific, and technological risks. It is costly and complex and has a track record of mismanagement. From the decision to embark on detailed design to the full experimentational operating of the facility may take more than 40 years. The science behind the facility can be questioned, it pushes technology boundaries, its economics are opaque and probably alarming to Treasury and Finance departments, and its political commitments and alliances are always exposed to geopolitical tensions.

Yet to some extent any assessment has to be calibrated against the scale of the potential of fusion energy and the opportunity cost of not doing it. The climate crisis and growing energy demands and insecurity requires a portfolio of energy provision to which fusion would contribute. Even if, worst case, ITER fails in its stated objectives, it still will have advanced technological and engineering competences around the world. It can also perhaps be calibrated against the fines paid following the financial crisis by JPMorgan Chase ($13 billion) and the Bank of America ($17 billion).[8] If these were the costs of dishonesty, what price can be put on the purpose of advancing science to produce clean energy?

When reflecting on ITER's contribution, Claessen (2020) refers to *technology diplomacy* and *diplomatic technology*. In the former view, technology has provided a valuable focus for collaboration between nations facing significant political tensions, such as the USA and USSR, and the USA and China. In the latter, ITER has brought together and developed the skills of scientists, engineers, policymakers, and administrators from many nationalities, through the common cause of developing fusion energy. It has also built the competences of the scientific and technological communities back home amongst contributing nations. As an example of a global

[8] https://www.marketwatch.com/story/banks-have-been-fined-a-staggering-243-billion-since-the-financial-crisis-2018-02-20.

effort at a time when globalization in many manifestations is being questioned, ITER represents an experiment for a better future that has engaged nations that account of half the world's population and 90 percent of the world economy.

A review by the European Commission cited by Claessens found the project will generate more value than it costs, and that companies working for ITER improved their capabilities and opportunities outside of fusion. The report found that as a result of their ITER contracts, 12% of industrial participants developed new cutting-edge technologies for use in areas other than fusion. Large science projects, such as ITER, develop construction and engineering capabilities and innovations in supply chains to deliver large and complex facilities and equipment. These extend through the industrial application of innovative technologies and intellectual property developed to undertake the science. Chapman, Cowley, and Wilson (2024) say that ITER has "driven substantial development of supply chains globally".

A 2021 article in the Financial Times article says ITER is a byword for glacial progress, but quotes Christofer Mowry, chief executive of the private sector company, General Fusion, saying:

> None of the private fusion companies would be here today without the science that was developed in the ITER programme," adding… "But the cost and timeline for ITER should not be used as a point of reference for what it takes to develop and commercialise fusion energy.[9]

One way to assess ITER is to simply to imagine walking into the centre of its tokamak and seeing the extraordinary knowledge and ingenuity it represents, admire its ambition, and congratulate ourselves that in our fractious and precarious world, humankind is still capable of attempting such marvels. For all its difficulties, it is attempting to build a vessel at such scale, and achieve reactions at such volumes, that have never been achieved before. Another way is to phlegmatically accept that ITER would inevitably face potentially insurmountable difficulties as a result of its Domestic Agencies putting national interests above that of the project. In line with our story of fusion energy progressing within innovation ecosystems from science to technology to commerce, the question is whether such a massive investment in what would essentially become a scientific project could more valuably be directed towards the application rather than the conduct of science. Such a judgement harshly ignores the relatively early stage of the development of fusion energy when ITER began, but also highlights the dangers of the sunk costs fallacy where organizations are reluctant to abandon or amend a course of action because so much has already been invested in it. ITER's contribution to future innovation ecosystems may lie in the serendipitous outcomes of experiments: in scientific discoveries, in the resolution of engineering challenges of large tokamaks such as the management of plasmas at large scale, in building industrial capabilities, and in demonstrating the continuing importance of international collaboration even under the greatest of

[9] "Nuclear fusion: why the race to harness the power of the Sunjust sped up", Wilson, T. and Bott, I, Financial Times, November 24th, 2021.

political strains. The question also needs to be posed of whether its efforts are being directed to an innovation ecosystem that has moved on.

6.2 The Princeton Plasma Physics Laboratory

The US Department of Energy's Princeton Plasma Physics Laboratory (PPPL) is one of the world's leading fusion research centres. It has an illustrious but chequered past, and ambitions to play a central role in the future development of fusion energy. Victor Hugo wrote that perseverance is the secret of all triumphs, and the PPPL offers a classic case of resolution and endurance driven by the promise of the conquest of viable fusion energy.

The Laboratory was founded by Lyman Spitzer (1914–1997) who played a significant role in early fusion research. Spitzer was a theoretical physicist and astronomer who studied at Yale, Cambridge, and Princeton before becoming Director of the Princeton University Observatory aged 33. While at Cambridge he was influenced by Arthur Eddington, the first scientist to suggest that the Sun shines by converting hydrogen atoms to helium. While at Princeton, where he was to spend the bulk of his career, he studied plasmas in interstellar space. Intrigued by the concept, if not the practice, of the fraudulent Argentinian experiment in fusion, he had the idea of confining plasma by a magnetic field in a machine he named the "Stellarator".

Spitzer's design for the Stellarator was approved for funding by the US Atomic Energy Commission, and magnetic confinement fusion research began at Princeton in 1951. In his proposal for research funding, he optimistically estimated that the Stellarator would produce around 150 MW of power, conceivably by the end of the decade, and by 1954 suggested that an investment of $200 million would produce a machine capable of producing thousands of megawatts (Siefe 2008).

The Princeton research was part of Project Matterhorn, a classified project related to the hydrogen bomb. The military connection—and research in this field did not become declassified until 1958—limited international exchanges of information, and publications, in the field at this time. Various experiments were conducted in the 1950s with different configurations of stellarator reactor design, including a figure-of-eight. In 1961, Project Matterhorn was renamed the Princeton Plasma Physics Laboratory, signifying the importance of its research into plasmas. Following confirmation of the performance of the tokamak in the USSR and, in 1969, the Laboratory's Stellarator was converted in short order into a tokamak. Six years later, the PPPL received Congressional approval to build a Tokamak Fusion Test Reactor (TFTR), aiming to be the world's first fusion device to experiment with a 50/50 mixture of deuterium and tritium. This reactor produced its first plasma in 1982.

As well as many significant scientific breakthroughs in plasma creation and containment, including the development of plasma thrusters and the propagation of intense beams of ions, the PPPL has registered some records in fusion. In 1978, its Large Torus machine achieved a temperature of 60 million degrees centigrade, the first time this temperature which is necessary for ignition in a tokamak had been achieved. In 1986, a temperature of 200 million degrees was produced; in 1990, this reached 400 million degrees, and in 1995, 510 million degrees. The TFTR recorded the production of 6.3 MW of fusion power in 1993 and 10.7 MW in 1994. It never, however, achieved its initial ambition of producing energy gain.

During the 1980s and early 1990s, PPPL ran a number of experimental machines and employed 1300 people. But it was facing straightened circumstances. Department of Energy funding for TFTR concluded in 1997. During this period, Congress slashed the research budget for fusion, and the decision was made in 2003 to continue to support ITER. In 2001, PPPL started work on the National Compact Stellarator Experiment (NCSX). By 2008, a year after it was supposed to be finished, with its cost tripling to $170 million, the Department of Energy cancelled the project, leaving PPPL with one machine, the National Spherical Torus Experiment (NSTX).

The NSTX was developed in collaboration with Oak Ridge National Laboratory, and Columbia and Washington universities. Design on this spherical tokamak began in 1995 and it created its first plasma in 1999. It was shut down in 2012 for a $94 million upgrade (NSTX-U), which was completed in 2015. The upgrade widened the central solenoid of the reactors, with the aim of increasing power, current, and pulse duration. After operating for 10 weeks, the NSTX-U was shut down because of poor performance of chilled copper wiring around the machine. Addressing this has required the complete dismantling of the reactor and coils, and the upgrade had yet to be completed a decade later.

These setbacks—the defunding of TFTR, cancelling of NCSX, and failure of NSTX-U—had inevitable consequences for the Laboratory. Staff numbers more than halved: in 2024, the PPPL employed around 500 researchers, engineers, and support staff. Tensions with the Department of Energy were exacerbated by Congress's fluctuating attitudes to fusion. The perception of poor performance was seen as damaging to Princeton University's reputation for excellence.[10] The experience of young people working at the Laboratory is decidedly mixed.[11] A 2010 external review of PPL's overall organization, management, and resources assigned to communications, outreach, and education made the telling recommendation that:

> Efforts should focus on articulating a strategic vision for the laboratory and, within the context of an overall strategic communications plan, sharing it with the laboratory community, stakeholders and patrons.[12]

The future looked more promising as fusion research gained more scientific and political support around the world. In 2019, the National Academies of Sciences,

---

[10] https://www.sciencemag.org/news/2020/02/after-decades-decline-us-national-fusion-lab-seeks-rebirth.

[11] Comments on Indeed.com in 2018/19 range from acknowledgement about the opportunities to work with 'the smartest people in the world' and 'one of the best opportunities I've had in my career' to 'not a good place to work', 'Average of staff is about 60 years old with a nasty attitude of entitlement...the culture is one of entitlement' and 'the management systems are hindered by the ancient ideals and slow pace of bureaucracy'.

[12] https://www.interactions.org/sites/default/files/2016-08/2010-PPPL-PeerReview.pdf.

Engineering, and Medicine (NASEM) recommended support for ITER and preparation for DEMO. It argued:

> now is the right time for the United States to develop plans to benefit from its investment in burning plasma research and take steps to develop fusion electricity for the nation's future energy needs. (2019:1).

It recommended that PPPL play a leading role in building a compact prototype energy plant using innovations such as powerful magnet coils made from high-temperature superconductors.[13]

Judgements on the performance of PPPL have to account for the related expertise; it has developed and the dual nature of its mission. By building and operating fusion devices, PPL claims to have gained extensive capabilities in advanced computational simulations and computer technology, vacuum technology, mechanics, materials science, electronics, and high-voltage power systems.[14] It has also been involved in building international capabilities through significant international collaborations, including with Germany's Max Planck Institute and South Korea's National Fusion Research Institute.

Its plasma research also extends beyond a focus on fusion energy. Its mission statement outlines how its research is "crucial to understanding the dynamics of the visible universe and solar system, and of high temperature in all contexts". It suggests this knowledge has a broad range of applications beyond fusion energy, including forecasting the impact of solar storms, semiconductor manufacturing, and the production of carbon nanotubes.[15]

The Laboratory hired Steven Cowley, previously CEO of the UKAEA, as its Director in 2018. A very eminent physicist, with a PhD from Princeton researching tokamaks, Cowley established an ambitious scientific agenda. He also aims to build stronger links between the Laboratory and the university and business, cementing its role as contributing to a fusion energy innovation ecosystem. He especially sees opportunities in applying its non-fusion research on plasmas in stars and interstellar space to process materials, particularly in semiconductors.[16] Its progress, however, will require secure and continuing commitments that the Laboratory has not enjoyed in the past. Such funding appears to be under threat in 2025 following the US government's change in funding for science, for example, in the associated overheads costs allowed on grants, and the PPPL responded with a hiring freeze and job losses.[17]

---

[13] National Academies of Sciences, Engineering and Medicine https://www.nap.edu/catalog/25331/final-report-of-the-committee-on-a-strategic-plan-for-us-burning-plasma-research.

[14] https://www.pppl.gov/about.

[15] https://www.pppl.gov/about/pppl-glance.

[16] https://www.sciencemag.org/news/2020/02/after-decades-decline-us-national-fusion-lab-seeks-rebirth.

[17] https://www.dailyprincetonian.com/article/2025/03/princeton-news-research-pppl-job-cuts-layoffs-graduate.

6.3 The UK Atomic Energy Authority (UKAEA)

During its history, which goes back to the end of the Second World War, the UKAEA has gone through many changes in its structure and remit affecting its scientific agenda and efforts to contribute to a fusion innovation ecosystem. Some of the major changes in the UKAEA are described in order to illustrate the challenging institutional framework in which fusion energy has developed in an international leader in fusion research.

6.3.1 A Complicated Organizational History

The UKAEA has experienced a number of major reorganizations, with significant changes in responsibility over time. Britain was the first country to split the atom (nuclear fission) in 1932, and it deepened its research in the ensuing years as other countries competed to develop potential civil and military uses of nuclear research. The UK Atomic Energy Research Establishment was established on 1 January 1946, 4 months after the conclusion of the Second World War. It was formed by the Ministry of Supply at RAF Harwell, 16 miles south of Oxford, and was the precursor to the UKAEA. The soon to be abandoned Royal Air Force airfield provided facilities and hangars suitable to accommodate construction of nuclear piles,[18] the reactors that would be built to test the development and control of nuclear power.

The requirement was for a site close to either Oxford or Cambridge Universities, which both had strong and growing capabilities in nuclear physics. RAF Harwell was one of several assessed for this purpose. Those around Cambridge, which had greater strength in nuclear physics, were unavailable as the RAF wished to retain its eastern airfields because of the emerging threat of the Cold War.

During the Second World War, Britain combined its nuclear science expertise with the USA in the Manhattan Project, but this collaboration was concluded by the Americans in 1946. Research in the field continued particularly strongly at the Clarendon Laboratory in Oxford and Imperial College. Demonstrating the UK's independent capabilities, two fission reactors were constructed at Harwell and these "went critical"—producing nuclear fuel sustained fission chain reactions—in 1946 and 1947.

By 1954, the UK's nuclear fission programme was 8 years old with the world's first demonstration reactor located at Harwell, fuel facilities at Springfields and Windscale, and a military nuclear deterrent programme at Aldermaston. The UK Atomic Energy Authority was formed on 19 July 1954 to bring these strands of nuclear R&D together. The UKAEA had a dual role: to develop reactor technologies for future nuclear power stations and to provide the UK's atomic weapons. Its

[18] Nuclear Pile is the name originally given to a Nuclear Reactor—a device used to initiate and control self-sustained nuclear chain reactions, required to create nuclear energy from fission.

mission was to carry out research, produce, use, and dispose of atomic energy and to make arrangements with universities and other organizations for research connected with atomic energy and radioactive substances. Its coat of arms represented both fission and fusion and its motto "E minimis – maxima": "From the smallest the greatest".

The Authority was already large in 1954, involving some 20,000 employees, operating from 10 major nuclear sites including Harwell, Windscale (renamed Sellafield in 1981) and Aldermaston. It grew and opened new sites during the 1950s. By 1956, the UKAEA had opened the world's first nuclear power station at Calder Hall in Cumbria—a Magnox reactor design—and with it, an ambitious plan to build 12 reactors. A fire at Windscale in 1957 set back plans, raising public concerns over the hazards of nuclear fission, which resulted in stringent safety improvements and regulation. Ten further Magnox power stations were constructed, including the Dounreay Fast Reactor, which began operating in 1959. In 1960s Britain, government policy pursued industrial development based on the "white-hot heat of technology", a phrase used by the UK's Minister for Industry, Tony Benn. This propelled investment in nuclear energy. By the early 1960s, the UKAEA employed more than 40,000 people in over 25 locations, and by the mid-1960s it had developed plans for new types of fission reactors including Advanced Gas-Cooled Reactors (AGRs) and Fast Reactors. By 1967, nuclear fission created more than 10% of Britain's electricity. Nuclear fusion research also continued with a purpose-built fusion laboratory opened at Culham in 1960.

Research and capabilities were developed adjacent to core nuclear programmes. Some of these were necessary to support the development of nuclear power. For example, in 1963, the UKAEA developed the X-ray cameras used on the UK's Skylark rockets in order to obtain high-quality images of the Sun's corona, providing more detailed knowledge of fusion temperatures.

The UKAEA underwent substantial restructuring following the growth and reorganization of the UK's nuclear industry in the early 1970s. Responsibility for fuel production was transferred to British Nuclear Fuels Ltd., isotope production to the RadioChemical Centre, weapons production to the Ministry of Defence, and radiological protection to the National Radiological Protection Board. In consequence the Authority became much smaller with around 13,000 employees operating in six main locations. It focused on research on civil nuclear power and expanding nuclear and non-nuclear commercial work.

From the mid-1990s onwards, the UKAEA focused on two activities: research into nuclear fusion at Culham, and managing and restoring the environment of its sites where fission research reactors and nuclear power plants had finished operating. This resulted in the development of significant capabilities in nuclear decommissioning, and, by 2005, 15 reactors had been safely dismantled. Land was released following the restoration programme, and some used for commercial development and as business centres for a variety of industries.

A separate Nuclear Decommissioning Authority (NDA) was formed in 2005 to take responsibility for the UK's civil nuclear clean-up programme. The UKAEA became a contractor to the NDA, managing Dounreay, Harwell, Windscale,

Winfrith, and JET facilities at Culham. At the same time, the UKAEA was also given permission by the Government to expand its work and compete for wider business in new areas.

Meanwhile, the UKAEA continued leading research on fusion at Culham, centred on JET. The Culham Centre for Fusion Energy (CCFE) was created in 2009 as a focus for its fusion research.

6.3.2 Commercializing Research: A Similarly Complicated History

Spinning-off UKAEA technologies into commercial and non-nuclear applications became a feature of its work from the mid-1960s, but its commercialization journey has been anything but smooth.

When in the 1970s, Mike Forrest—having returned from his successful time in Moscow—was told his duties at Culham were to extend to take in commercialization, he thought this was going to be a boring option, but he was pleasantly surprised by the opportunities it opened. He reports trips to Japan and Libya, with the latter producing a contract for £2 million. By 1985, Harwell was generating £50 million per year from more than 1200 contracts in areas as diverse as space tribology, petroleum services, and reactor safety testing. The changing nature of its work resulted in the UKAEA becoming a "trading fund", with the autonomy to manage its income and expenditure separately from detailed government departmental scrutiny. Growth in commercial activities led to the formation of a separate commercial arm, AEA Technology, in 1989.

During the 1980s and early 1990s, the UK Government's civil fission nuclear research programme shrank and the civil nuclear programme ended with the closure of Dounreay in 1994. At the same time, non-nuclear and nuclear commercial work had grown significantly, and AEA Technology was spun out in 1994 and privatized in 1996, sold for £228 million. Its share price doubled in 2 years, and the privatization was criticized for under-estimating its value. Despite AEA Technology diversifying into general engineering and environmental consulting, it struggled after a period and was sold to engineering consultants Ricardo in 2012 for £18 million.[19]

A spate of commercial developments and re-organizations ensued and in April 2008, UKAEA Ltd. was formed as a commercial business in nuclear site decommissioning. UKAEA Ltd. was sold to Babcock International Group plc 18 months later. The UKAEA continues to own the freehold of the majority of the Harwell Campus and the Culham Science Centre. The nuclear licensed part of the site at Harwell is leased to the NDA which is responsible for overseeing its decommissioning. In 2008, the Authority formed a joint venture to develop the Harwell Science and Innovation Centre into a site for science, innovation, and technology.

[19] https://www.ft.com/content/6e26154e-29c3-11e2-9a46-00144feabdc0.

6.3.3 The UKAEA Today

In it 2024/25 Annual Report and Accounts (UKAEA 2025), the UKAEA defines its mission as "Leading the development of sustainable fusion energy and maximizing the scientific and economic benefit". A public-sector research organization, it is classed as a non-departmental public body sponsored by the Department for Energy Security and Net Zero. It defines what it does as: building the knowledge base of fusion; delivering fusion powerplants, systems, and technologies; and enabling the fusion community (UKAEA 2025).

The UKAEA is operationally complex in a number of senses for an organization of around 2500 employees. This includes the complexity of the science which requires the integration of different knowledge bases, often at the global forefront of understanding, and the further integration of that knowledge with practical technologies and engineering. It also involves the explorations of experiments with the demands of delivery. There is a Project and Programme Management Office to assist with delivery, and a stage-gate system is used to filter and manage projects.

Key research areas at the UKAEA include testing materials and components in extreme conditions, controlling extreme heat and extracting it without damaging components, maintaining and replacing components robotically, and using advanced computer simulation to bridge the gap from experiment to power plants with advanced simulation. To do this, as well as running the decommissioning of JET, the UKAEA also includes a number of complementary organizations. They include: MAST-U, dedicated to managing heat exhaust; H3AT, for use, storage, and transport of tritium; RACE, for robotic handling; MRF, the Materials Research Facility; and FTF, the Fusion Technologies Test Facilities. The Authority also possesses significant advanced computing and digital design capabilities. All these capabilities are informing the creation of STEP.

Aware that STEP will be employing people who are graduating today, attention is paid to recruiting and developing fusion researchers and engineers. There are connections with 36 Universities, and the Authority helped create and sponsors Professorial Chairs, such as the UKAEA Chair of Digital Engineering for Fusion Energy at Manchester University.

6.4 Research Institutes and the Building of Industrial Capacity

The benefits for industry of fusion research in places such as the UKAEA, PPPL, and ITER appear *upstream* in developing the construction and engineering capabilities and innovations in supply chains to deliver large and complex facilities and equipment. These capabilities are of great value to the innovation ecosystem, including to private sector fusion firms. They extend *downstream* through the industrial application of innovative technologies developed to undertake the science. The

World Wide Web and the laser are classic examples of tools developed to help undertake scientific experiments. Some of these benefits are planned and expected, some occur through serendipity, such as no-one foresaw the industrial applications of the web or lasers. Furthermore, valuable skills are developed in demanding science projects that have potential use in upstream and downstream industry. One of the major issues for some firms working with fusion is the need to operate in radioactive environments, albeit different to fission but still with their specific demanding regulatory requirements.

There is some evidence that manufacturing and constructing ITER's facilities has enhanced industrial capabilities and stimulated innovations upstream. The review by the European Commission mentioned earlier found the project will generate more value than it costs, and that companies working for ITER improved their capabilities and opportunities outside of fusion. The project has involved significant advances in construction technology, and guarantees in the contracts in science projects can reduce the risks for industry partners investing in their own R&D. The UKAEA had a policy with JET of procuring functional rather than detailed designs, allowing industrial suppliers to develop and exercise their capacity to innovate.

One of the two objectives of STEP is to build industrial capabilities. The intent is to involve partner-suppliers who share a concern for innovation, with the aim growing capabilities in supply chains for fusion and create spin-offs in adjacent industries. Fusion research at the UKAEA has already provided opportunities downstream for new small firms. Luffy AI is a start-up based at Culham and uses adaptive intelligence to improve industrial controls.[20] Oxa is a related Oxford University spin-off, based at Culham, that is developing autonomous vehicle software using 10 km of on-site private roads to carry out vehicle testing. Oxa uses computer vision, machine learning, and artificial intelligence to enable vehicles to operate autonomously in any environment in sectors including aerospace, automotive, construction, logistics, and mining.[21]

The severe demands of the science, with vast heat and material changes in fusion vessels, have provided extensive opportunities for industry. RACE, the Remote Applications in Challenging Environments programme, which supports and sits alongside JET, is developing remote handling robotic technologies that can be used in a variety of extreme conditions found in industries including nuclear, petrochemicals, space, construction, and mining.[22]

The remote handling system developed for JET has two planar robotic arms. Robots are needed to deal with parts of the fusion vessel that had to be inspected, maintained, and upgraded. RACE has patented a "robotic snake" that can operate remotely on pipework in hazardous conditions. Using an ultrasonic sensor system, it can direct a laser welding tool in conditions where pipes are packed tightly together. The same principles are being applied to a laser cutting tool. Rob

[20] https://luffy.ai/#home.

[21] https://www.oxbotica.com/.

[22] https://race.ukaea.uk/about-race/.

Buckingham, the Director of RACE, explains the importance of robots for keeping fusion plants continuing to operate. Fusion plants, he says, have to be designed for ease of use and preferably use robots for manufacture, assembly, operations, and maintenance. RACE focuses on four areas: in-vessel, ex-vessel, transport, and the robotic "nervous system". Buckingham says that as so much attention in fusion is directed at plasmas in vessels, there is the need to focus even more on the latter two issues.

RACE is one of the more externally focused elements of the UKAEA, with strong connections into the wider robotics community. It also has around 20 academic partnerships, including the Oxford University Robotics Institute.

Since his appointment in 2018 as UKAEA Chief Technology Officer, Tim Bestwick, has been working to develop a more sophisticated programme for managing intellectual property. Conscious of the role of public research organizations in producing freely available, public good information, he also points to the value of patent protection not only in fusion machines but in related and adjacent industries. In his view, the key areas for commercialization include robotics, diagnostics including sensors, tritium handling, controlling plasmas, high superconducting magnets, and potentially in materials, although these may be specific to fusion as they need to deal with neutron flux. Bestwick was appointed Deputy CEO of the UKAEA in 2024 and Interim CEO in 2025.

Both upstream and downstream innovations are derived from investments *instream*, within projects necessary for the conduct of scientific experiments. Scientific projects that have the most demanding new requirements in order to undertake experiments can have the highest potential for upstream and downstream benefits. In the long term, this means that industry can benefit most when scientific projects are at their most ambitious (Dodgson and Gann 2023). Ian Chapman, CEO of UKAEA from 2016 to 2025, relates how a very demanding technical problem at Culham of winding coils on powerful magnets in completely sterile conditions led a small UK MRI-manufacturing company to conduct new experiments to develop a solution. They now use capabilities they developed in their other product ranges, increasing sales by 30 percent.

That is not to say that less ambitious scientific projects aren't of value to industry, nor that science projects shouldn't be very cautious about managing their ambitions and costs. Upstream and downstream benefits can be planned and encouraged in all cases, recognizing that upstream benefits can be realized relatively quickly: they need to deliver something known and planned; while downstream benefits can be more speculative.

It is just that when scientists and their support teams design experiments to address the most difficult and demanding scientific problems, their depth of knowledge and ingenuity will produce highly imaginative, innovative, and practical solutions. Industry has the opportunity to benefit from these solutions, which would not emerge without investment in basic science and accessing the brainpower and skills scientists and their support teams possess. This emphasizes how interactions in a fusion energy innovation ecosystem is not simply single directional. Research does not produce pieces of results that are once and for all handed over to industry: as shown by Nathan Rosenberg (1982), continuous interactions in all directions are needed, as seen in the STEP Programme and the demands it will make of the UKAEA.

6.5 The Challenge of Tritium

One area where the connections between research and industry will be absolutely crucial is the development of tritium. Although other fuels are being explored in fusion experiments, the combination of deuterium and tritium is favoured in tokamaks. While deuterium can be used by itself in creating plasma pulses, commercial operation necessitates its 50/50 mixture with tritium to produce continuous operation of plasmas.

The availability of tritium presents one of the starkest challenges facing fusion energy (Jassby 2024). To describe tritium as a difficult material would be an understatement: it is aggressive, corrosive, volatile, explosive, flammable, and embrittles many materials it comes into contact with. It presently needs to be manufactured in nuclear reactors or accelerators, and one estimate suggests that in 2024 there was no more than 30 kg of tritium in the world.[23] Given its rarity, and hence very high cost, the optimal way of proceeding is for future deuterium-tritium fusion machines to create their own fuel. This will require the production, capture and use of tritium in the containing "breeder blanket" of the vessel, which is likely to be made of lithium.

The benefit of using tritium lies in how enormous amounts of energy can be produced from miniscule amounts of it. In the 30 odd years that JET conducted its experiments, it used less than 100 g of tritium, around the weight of two medium-sized eggs. The world record creating JET experiment in 1997 used 20 g of tritium, and while the experiment was a great success, it was learned that around 70 percent of the tritium was absorbed by the vessel walls, which in those days were carbon-based. Such a loss would be totally uneconomic and, in 2011, the tiles of the wall were replaced with Beryllium, which absorbs about 10 percent of the tritium collected in the old carbon walls. Beryllium is a difficult material in that it is fragile, and as a metal it melts. It is also highly toxic and its dust is carcinogenic, so dealing with it safely is a major consideration.[24] Because of these difficulties, ITER is now designed to use tungsten.

The production of tritium is confronted by the fact that, as a form of hydrogen, it can permeate material even as solid as stainless steel, and that it is radioactive. The challenge is to construct a facility from which no tritium escapes, which is especially difficult given the high heat it is exposed to in fusion reactions. A solution is to surround the vessel with nitrogen as a secondary means of containment to prevent the escape of neutrons.

[23] See Jassby (2024).

[24] See e.g. https://www.iter.org/newsline/-/2763.

There is presently around 60 g of tritium at Culham, with plans to have around a kilo. ITER will require 6 kilos. Damian Brennan, Head of H3AT (Tritium Advanced Technology) at Culham until 2023, states that fusion will fail economically as a technology if it uses the same methods of disposing radioactive waste from fission (essentially burying it deep in the ground in concrete). As JET is decommissioned, however, some of its radioactive surfaces and materials will be removed to Harwell for processing before being buried in the ground. Much radioactive waste—around 30 tons up to 2022—has been processed at Culham, reducing its radioactivity such that is it useable in medicine. The design of H3AT is entertainingly described by Brennan as involving a "whacking great pizza oven producing temperatures of a 1000 degrees centigrade, tied alongside a 'Mercedes Benz' catalyst, tied along three beer barrels of water". The objective is for tritium to be captured in the barrels and taken back to the plant for re-use. H3AT plans to produce 100 g of Tritium. The approach has been to design the system for minimal tritium loss.

Industry is centrally involved in the development of H3AT, and all that development is "ITER-relevant", is useful for the fusion industry more broadly, and is particularly directed towards the design and construction of STEP. The aim is for the UKAEA to progress from being a client to industry to selling knowledge to it, so it can win contracts. It is also believed that Culham's expertise in de-radiating materials from fusion can be applied to those produced from fission. As a result of a collaboration with Bristol University, it has been found that some decommissioned material can be transformed into diamonds that include tritium in them for use in specialist batteries, for example in pace-makers, and a spin-out company, Arkenlight, has been formed to commercialize the technology.[25]

In March 2025, the Italian energy company, ENI, announced a collaboration with UKAEA to construct the world's largest and most advanced tritium fuel cycle facility the "UKAEA-Eni H3AT Tritium Loop Facility". Located at Culham Campus, it is planned to be completed in 2028. According to the company, Eni will contribute its expertise in managing and developing large-scale projects and its industrial-scale capabilities in plant engineering, commissioning, and operations, helping to de-risk the project.[26]

Also in 2025, the UKAEA announced a £200 million Lithium Breeding Tritium Innovation (LIBRTI) programme to demonstrate controlled tritium breeding. Over its 4-year span, the programme will purchase a neutron source as the backbone of what is believed to be a first-of-a-kind testbed facility. The UKAEA also aims to offer £9 million funding for 12 small-scale tritium breeding and digital simulation experiments involving over 20 small firms and universities. The objective of the programme, which is crucial for STEP, is to breed tritium in a lithium-containing blanket which also absorbs heat and shields radioactivity. An in-house tritium

[25] https://www.arkenlight.co.uk/.

[26] https://www.eni.com/en-IT/media/press-release/2025/03/fusion-energy-eni-and-ukaea-to-build-tritium-fuel-cycle-facility.html.

training course is planned to be offered to industry, and the UKAEA chairs the UK Tritium Group.

6.6 Other Institutions

Fusion research institutes are, of course, to be found in other nations around the world. Germany's Wendelstein 7-X, for example, operated by the Max Planck Institute for Plasma Physics, is the world's largest and most advanced stellarator, and is working with start-up companies to further develop the technology.[27] Korea's Institute of Fusion Energy operates KSTAR (Korea Superconducting Tokamak Advanced Research), which is one of the first tokamaks to feature fully superconducting magnets.[28] The French Alternative Energies and Atomic Energy Commission (CEA) operates a tokamak named WEST, which is achieving records in plasma duration.[29] China's increasingly important fusion research institutions will be discussed in the following chapter.

These, and other institutions in other countries, operate research machines with research objectives. They are all connected with national and international research efforts; all adding valuable knowledge, and developing important capabilities in research and industry, vital to the production of viable fusion energy. While other innovation ecosystems can be initiated by industry, for example, in pharmaceuticals and electronics, fusion energy innovation systems emerged from, and still heavily rely upon, the contributions of research institutions.

References

L. Bardin, (2020) "How do organizations improve cross-border partnerships", https://discovery.ucl.ac.uk/id/eprint/10093270/

I. Chapman, S. Cowley, H. Wilson, The spherical tokamak for energy production: Theme issue introduction. Phil. Trans. R. Soc. A **382**, 20230416 (2024)

M. Claessens, *ITER: The Giant Fusion Reactor* (Springer Nature Switzerland, 2020)

A. Davies, M. Dodgson, D. Gann, and MacAulay S. Bringing innovation into large, high risk projects, MIT Sloan Management REview, **59**,1:73–78, (2017)

M. Dodgson, *Technological Collaboration in Industry: Strategy Policy and Internationalization in Innovation* (Routledge, London, 1993/2018)

M. Dodgson, D. Gann, Innovation and big science projects, in *Handbook on Innovation and Project Management*, ed. by A. Davies, S. Lenfle, C. Loch, C. Midler, (Edward Elgar, Cheltenham, 2023), pp. 423–434

[27] "German start-up wins initial funding for revolutionary fusion energy machine" Financial Times: https://www.ft.com/content/c777f0db-1a4b-4ced-b02a-c27178032a50.

[28] https://www.kfe.re.kr/eng.

[29] https://www.cea.fr/english/Pages/News/nuclear-fusion-west-beats-the-world-record-for-plasma-duration.aspx.

B. Flyvbjerg, N. Bruzelius, W. Rothengatter, *Megaprojects and Risk: An Anatomy of Ambition* (Cambridge University Press, Cambridge, 2003)

D. Jassby, The fuel supply quandary of fusion power reactors. Bull. At. Sci. **80**(6), 377–384 (2024)

R. Khatchadourian, *A Star in a bottle* (New Yorker, 2014)

National Academies of Sciences, Engineering, and Medicine, *Final Report of the Committee on a Strategic Plan for U.S. Burning Plasma Research* (NASEM, Washington, 2019)

N. Rosenberg, *Inside the Black Box: Technology and Economics* (Cambridge University Press, Cambridge, 1982)

C. Siefe, *Sun in a Bottle: The Strange History of Fusion and the Science of Wishful Thinking* (Penguin, New York, 2008)

UKAEA (2025), Annual report and accounts 2024/2025, presented to parliament pursuant to sections 3(5) and 4(3) of the atomic energy act 1954, ordered to be printed by the house of commons, 17th July 2025

Chapter 7
Key Players in Fusion Innovation Ecosystems: Government and Industry

7.1 Governments

The large government investments in fusion research in the world after the Second World War was initially attracted by military applications and latterly also by its potential for energy generation. These latter investments have been made in the full understanding that this potential will only materialize many decades after the politicians and policymakers who support them will be retired or dead.

Chapman, Cowley, and Wilson (2024:3) provide a useful summary of government strategies for fusion energy:

> The most established public programmes are the coordinated European effort to design a demonstration power plant, DEMO; the UK fusion strategy including a unique approach to a public–private partnership; the US National Academies report, which stimulated a White House 'bold decadal vision' for fusion and the subsequent public–private partnership programme; and the Chinese Fusion Engineering Test Reactor now with a broad range of participants under state control. All of these programmes aim for prototype power plants to commence operations before the middle of the century. Beyond these well-established strategies, other countries have indicated increased support for fusion development, especially through public–private partnerships, such as in Germany and Japan.

To briefly encapsulate the journey government policy has made: to begin, the focus was on science policy, which due to the cost and complexity of research facilities, such as ITER and JET, were international efforts. Latterly, as the science became better understood, there has been greater focus on national schemes, with greater attention on industrial strategies for developing and leveraging the capabilities needed to build fusion machines and eventually a fusion industry. While there has been fluctuations in policy enthusiasm for fusion in the USA, in the UK, and, despite sometimes high levels of political volatility, there has generally been consistent support for the development of fusion science, technology, and industry.

M. Dodgson, D. Gann, *The Search for Viable Fusion Energy*, SpringerBriefs in Applied Sciences and Technology, https://doi.org/10.1007/978-3-032-13466-0_7

Notably, the development of fusion energy has received significant support from recent Prime Ministers.

The balance between government and industry involvement reflects political choices: in China, the key player is inevitably the government; in the UK, government is the major actor although there is a move towards more private sector engagement, for example, through the public-private partnership, STEP; in the USA, government has a key role although, given the emphasis placed on, and success of, US technological entrepreneurship, much is expected of the private sector, albeit alongside a range of US government schemes designed to encourage private sector investment. Along with the thesis of this book, fusion innovation ecosystems benefit from inputs from both government and the private sector, and questions can be raised on whether government-led approaches without private sector profit motives, or private sector-led approaches without the resources of government, will produce the most beneficial results. The future of viable fusion energy depends in many ways on learning lessons over time from the strengths and drawback of these different approaches, with the hope eventually to discover an optimal balance.

7.1.1 The UK

The UK government was the first to develop a specific strategy for fusion, launching its Fusion Strategy: "Towards Fusion Energy", in October 2021.[1] This outlined its ambitions to:

> lead the development of this low carbon energy technology and secure the economic opportunities of a future global fusion energy market.

Its goals include the:

> UK demonstrating the commercial viability of fusion by building a prototype fusion energy plant in the UK that puts energy on the grid, and for the UK to build a world-leading fusion industry which can export fusion technology around the world.

The UK also became the first country in the world to legislate for fusion energy, which provides clarity on the regulatory regime for fusion energy facilities and helps address some of the uncertainty for the fusion industry. Significantly, Parliament legislated that fusion power plants would not be regulated as nuclear installations. This means they do not need to be located on previously approved nuclear sites. Within the industry, it is argued that regulation proportional to the risk from fusion means that the pathway to commercialization became much clearer, with much lower costs than would otherwise have been the case.

---

[1] https://www.gov.uk/government/publications/towards-fusion-energy-the-uk-fusion-strategy.

The UK Government updated its Fusion Strategy in October 2023.[2] It outlines a Fusion Futures programme focusing on three core areas: infrastructure, including funding of facilities for fusion companies and innovation in fusion fuels generation; skills, creating a new Fusion Skills Centre to work with universities, colleges, and employers, to provide a pipeline from apprentices to post-doctoral fellows; and, industrial and commercial opportunities, including working with ITER and exploring the development of a UK fusion investment fund. The Strategy promised—in the light of the UK's disassociation from ITER—up to £650 m of new investment until 2027, subject to business case approval, on top of existing fusion programmes. In June 2025, the UK government announced a record £2.5 billion to develop a prototype fusion energy plant.[3]

7.1.2 The USA

Government policy developments in the UK helped inform similar movement elsewhere. In March 2022, the US Government produced its fusion policy: "Developing a Bold Vision for Commercial Fusion Energy".[4] The Statement refers to the need for fusion power and its potential as a "game changing" clean energy technology, the strength of US science in the field, and the increasing activity in the private sector. It offers a decadal vision to accelerate fusion power, including launching an agency-wide initiative to accelerate the viability of commercial fusion energy in partnership with the private sector. The US Government updated its Fusion Energy Strategy in 2024,[5] following further scientific success, particularly the fusion ignition achievement at the LLNL, described earlier.[6]

Based on the demonstrable success of its model in numerous technologies, the US's approach emphasizes the central role of the private sector, well connected into the strengths of its research base, and supported by substantial pools of venture finance and risk-tolerant entrepreneurs. While it is relatively small scale, the US also provides an example of the way governments can demonstrate and de-risk speculative and adventurous scientific investments. In 2007, the US government launched the Advanced Research Projects Agency–Energy (ARPA-E) programme, under the Department of Energy. The programme was designed to target promising and underexplored approaches to the development of advanced energy technologies. ARPA's overall aim is to:

---

[2] Chrome extension://efaidnbmnnnibpcajpcglclefindmkaj/https://assets.publishing.service.gov.uk/media/65301b78d06662000d1b7d0f/towards-fusion-energy-strategy-2023-update.

[3] https://step.ukaea.uk/2-5-billion-for-world-first-prototype-fusion-energy-plant/.

[4] https://bidenwhitehouse.archives.gov/ostp/news-updates/2022/03/15/fact-sheet-developing-a-bold-vision-for-commercial-fusion-energy/.

[5] chrome-extension://efaidnbmnnnibpcajpcglclefindmkaj/https://www.energy.gov/sites/default/files/2024-06/fusion-energy-strategy-2024.pdf.

[6] https://lasers.llnl.gov/science/achieving-fusion-ignition.

> identify and promote revolutionary advances in fundamental and applied science, translate discoveries and inventions into innovations, and accelerate development of uncertain and unproven ideas into the private sector.

Essentially it aimed to fund high-risk, high-reward ideas where there is a high chance of failure. In 2014, ARPA-E launched its ALPHA (Accelerating Low-cost Plasma Heating and Assembly) project to develop and build prototype tools that demonstrate new methods in fusion and that build foundations for future reactor designs. This was ARPA's first investments in fusion and ALPHA supported nine projects exploring intermediate density fusion, in technologies such as sheared flow Z-pinch. Support was provided to universities, government laboratories, and private sector firms. A review of the programme found a number of future possibilities, but none that performed at a higher level than existing tokamak and magnetic ignition technologies.[7] The review pointed out that the purpose of the scheme was exploratory and was not designed to produce prototypes, and it recommended a follow-up programme.

In July 2020, the Department of Energy announced $32 million in funding for 18 projects as part of the Breakthroughs Enabling THermonuclear-fusion Energy (BETHE) program.[8] BETHE projects build on ALPHA to grow the number of privately funded fusion companies. The objectives of BETHE research is to reduce cost, unit size, and complexity of fusion systems, and smooth the path to fusion commercialization to include public, private, and philanthropic partnerships.

The future of these schemes will depend on the US government's approach to energy research and to science, post 2025.

7.1.3 China

China is undoubtedly a key player in the pursuit of fusion energy. Its approach to-date has been multi-staged, with a long-term commitment to ITER, from which it has learned much, and to which it has provided research results and components including vacuum vessel modules, and the clearly articulated aim of advancing domestic energy security and positioning the country as a key player in the global quest for sustainable energy. Each stage aims to build upon the lessons of its predecessor, and its coordinated, long-term effort signals China's ambition to produce controllable fusion energy by the middle of the twenty-first century. This marks a shift from international "catch-up" to "co-leadership", characteristic of China's overall development path in science and innovation (Zhang, Dodgson, and Gann, 2022).

---

[7] Retrospective of the ARPA-E ALPHA fusion program, Nehl, C., Umstattd, R., Regan, W., Hsu, S., and McGrath, P. (2019). https://ui.adsabs.harvard.edu/abs/2019arXiv190709921N/abstract.

[8] https://arpa-e.energy.gov/programs-and-initiatives/view-all-programs/bethe.

Much of the coordination has been facilitated by the state-owned China National Nuclear Corporation (CNNC) and its venture, the CNNC Fusion Technology Institute. The CNNC co-founded the China Fusion Energy Consortium with 25 state-owned enterprises and universities working to advance fusion reactor engineering. Another key player is the Institute of Plasma Physics (IPP) at the Hefei Institutes of Physical Science (HFIPS), part of the Chinese Academy of Sciences.

The first stage, the EAST project located at HFIPS, has been operational since 2006. EAST is the world's first fully superconducting tokamak and has set multiple world records for plasma duration and temperature. These achievements have provided experimental data for ITER and laid the groundwork for future reactors by validating key technologies such as its magnets and advanced plasma heating methods. This stage has essentially involved exploring the physical limits of plasma confinement and stability.

Launched in 2024, and operational in 2025, the second stage, the Comprehensive Research Facility for Fusion Reactor Key Systems (BEST), is an engineering testbed for China's next-generation reactors. Also based at the HFIPS, its research targets include the development of advanced divertor systems to handle intense heat and impurity flows, breeding blankets for tritium fuel self-sufficiency, and the integration of superconducting magnets with reactor-scale vacuum systems. Although a research facility, BEST is focused on addressing the practical challenges of building a working fusion power plant. For example, it completed a 1/8 scale prototype of a vacuum chamber for remote maintenance testing. This stage aims at developing and validating the critical engineering systems required for reactor operations.

Led by CNNC, the China Fusion Engineering Test Reactor (CFETR) is China's flagship fusion engineering project, positioned as its next major step. Planned for completion by 2035, with demonstration power generation targeted for 2050, CFETR is being designed to achieve long-duration reactions, aimed at gaining ten times the energy used to produce fusion, and deliver 200 megawatts of net electricity to the grid. The reactor aims to validate a closed fuel cycle with tritium breeding and bridge the gap between international experiments such as ITER and future commercial fusion plants. This stage aims to be a prototype for commercial fusion energy.

China is evolving its fusion innovation ecosystem by building on government-led R&D and significant expertise in research institutes, and encouraging market-driven initiatives. A number of private sector startups have emerged, including STARTORUS Fusion, Energy Singularity, ENN Fusion, and CAS Hiao, offering design and manufacturing expertise, and providing specialist contributions to supply chains. We speculate further on the importance of China for viable fusion energy in Chap. 8.

Other governments will pursue different balances of approach, but China, which is predominantly government-led; the USA, with its inclination towards the entrepreneurial private sector; and the UK and its search for private-public sector partnerships, offer a highly idealized spectrum of approaches that governments can take to further the development of fusion energy.

7.2 Industry

In an ideal conceptualization of a fusion energy innovation ecosystem, the pursuit of fusion energy builds on the scale and long-termism of government investment, backed by effective regulation, complemented by the speed, agility, risk tolerance, and market discipline of the private sector. Scientific experiments funded by governments reduce uncertainties and help reduce risks for private sector investments (the "incomplete" and "asymmetrical" information issues described by innovation theory). A number of large companies, including Lockheed Martin and Toyota, have invested to develop fusion energy commercially, which has also attracted around 40 small "fusion" firms. These small fusion firms are optimistic about producing fusion energy commercially ahead of government laboratories. A survey by *Nature* suggests two-thirds of them believe a fusion plant will be delivering electricity by 2035.[9] Zhong (2024) reports:

> They are signing presale deals with customers, developing supply chains, cultivating a work force, talking with regulators — all the elements that will be needed to make fusion an affordable, practical power source, not just a science experiment.

The question remains of the extent to which industry has been committed and is prepared to dedicate resources to the fusion industry at the level required to make fusion energy viable.

There are a number of shared features of small fusion companies. First, they have commonly spun-off from universities and research facilities and retain strong links with them. Second, despite having very impressive advisors and investors, their claims on what they can achieve are often over-ambitious. Third, they are experimenting with a diverse range of technologies, very important for the development of the field and the search for dominant designs, and are, for example, making significant contributions in the management and use of data. Fourth, some are developing impressive capabilities as suppliers of specific subsystems of technologies and services. Fifth, given the complexity and expense involved in commercializing fusion energy, their future will depend on the nature of their connections with government and other elements in fusion innovation ecosystems.

Universities and research institutes have commonly provided both the seedcorn for these small firms and a source of continuing support. TAE Technologies, based in California, emerged from the University of California; Commonwealth Fusion Systems, based in Massachusetts, was formed from MIT and retains very close connections with it; the Canadian firm, General Fusion, was started by a scientist from the University of British Columbia; and First Light Fusion, based in Oxfordshire, spun out from Oxford University. Tokamak Energy is a spin-out from the UKAEA, and Zap Energy, based in Washington State, is based on research at the University of Washington and Lawrence Livermore Laboratory.

[9] https://www.nature.com/articles/d41586-024-02247-2#:~:text=In%20a%202023%20survey%2C%2065,by%202040.

As Clyne (2020: np) puts it:

> One advantage of the newer, small-scale fusion projects is that they can concentrate on the novel aspects of their designs, while taking advantage of decades of hard-won knowledge about the fundamentals of fusion science.

Connections with universities and research organizations remain crucial for these small firms. TAE enjoys many collaborative research projects with universities, including Princeton, University of California Irvine and Los Angeles, and Wisconsin Madison. It collaborates with the Lawrence Livermore Laboratory and one Italian and one Japanese university. First Light Fusion collaborates closely with a number of academic organisations, including the University of Oxford, Warwick University, UCL, and Imperial College London.

The list of high-profile billionaires who have invested in fusion companies include: Bill Gates, George Soros, Jack Ma, Richard Branson, Peter Thiel, Jeff Bezos, Paul Allen, and Sam Altman; and they have invested in companies that commonly have very eminent advisors. The Board of Directors of TAE Technology, for example, includes a former US Secretary of Energy, the former CEO and Chair of GE, and former Chair and CEO of Morgan Stanley. The Advisory Board of First Light Fusion also includes a previous US Secretary of Energy (and Nobel Prize winner), and a previous UK Chief Scientist. A number of large companies and sovereign wealth funds have also invested in these firms, including Eni, Chevron, and Google, and the Malaysian and Norwegian governments. They have also in many cases benefitted directly from government schemes aimed at encouraging fusion energy: overall these companies received investment by 2024 of around $9 billion, with a small number of them now being valued in billions of dollars. In 2021 alone, for example, Commonwealth Fusion Systems received an investment of $1.8 billion, and Helion of $500 million.

The early promises of many of these companies—commonly that they would be delivering viable fusion energy by the mid-2020s[10]—has proven unfounded as none have done so. So the question is why have such successful and eminent people become involved with them? According to Coopersmith (2024):

> They have billions of personal dollars on hand—so they can take part of that money out and afford to lose it. But if one of their bets on fusion pans out—even years down the road—it will deliver a 1000 or a 10,000 percent return on investment, a "super return." So they place a few bets. Or lots of bets.

First and foremost has to be the overall potential of fusion as a carbon-free source of energy: they are buying into the option of being part of a revolution in energy production.[11] Other reasons include their entrepreneurship and the novelty of the technology they are developing.

In a study of one startup, Tokamak Energy, Pearson et al. (2020) argue that such firms offer an alternative model of innovation in fusion to publicly funded

---

[10] See Clyne (2020).

[11] See, for example, Bill Gates's view on the globally transformative potential of fusion: https://www.gatesnotes.com/the-future-of-energy-is-subatomic.

laboratories. In contrast to the latter, which are argued to follow a linear path of development with little concern for innovation and commercialization, startups are argued to pursue "agile innovation", where technology is developed flexibly and iteratively towards explicit commercial goals. Such a distinction completely fails to capture the reality of innovation in fusion. It fails, for example, to appreciate the commercial applications that have emerged from public sector laboratories, the crucial role they play in producing the fundamental science and trained scientists that feed startups, and the safety and regulatory concerns that public laboratories address to the benefit of the industry. It also ignores the fundamental scale of the challenge of developing fusion plants, which are way beyond the capacities of small firms. As Pearson et al. acknowledge, when scientific problems emerge in companies, such as the one they studied, they often turn to public laboratories for solutions. In short, it is not a case of startups being an alternative model of innovation, but a complementary one in an innovation ecosystem.

Successful innovation ecosystems, like ecosystems found in nature, encompass diversity, which, in the case of fusion, is seen in the different scientific and business experiments being undertaken in these startups. Tokamak Energy, for example, is pursuing small spherical tokamaks, taking advantage of progress in high-temperature superconductor materials, and is making advances in areas such as quench protecting superconducting magnets by dissipating heat by means of insulating the magnetic coils. TAE Technologies uses hydrogen-boron as an alternative to deuterium-tritium in what it calls aneutronic fusion. This requires significantly hotter temperatures than other fusion machines, but the company contends that hydrogen-boron is the safest fuel cycle with lower environmental impact, particulate emissions, and radioactivity. General Fusion is attempting to combine both magnetic and inertial approaches in what it calls magnetized target fusion. Zap Energy, based in Seattle, is developing a fusion machine that stabilizes plasma with a sheared flow—plasma flowing at different velocities at different radii—which it argues can create the high-temperature, high-density reactive medium confined long enough for fusion reactions to occur. Helion, based in Washington State, uses an approach called pulsed non-ignition fusion, involving a dumbbell shaped vessel 40 ft. by 6 ft., accelerating plasma at 100 million degrees so that the reaction pushes back on the system's magnetic field.

In an extraordinary example of technology emulating nature, First Light Fusion is basing its approach on the PhD of Nick Hawker, one of its founders, in which he studied the claw of a pistol shrimp. This remarkable crustacean stuns its prey by firing its claw, and in doing so creates shockwaves so powerful that they compress vapour in the prey analogous to a natural form of inertial confinement fusion. The company's research is trying to mimic the shrimp's physiology and use advanced implosion processes to achieve high temperatures and compressions; calling its approach "projectile fusion". It's research electro-magnetically launches a copper disk at high velocity, at more than 20 km a second, to collide with a target, a pellet of fuel with a hydrogen bubble inside. It is studying the use of pressure from intense shock waves to crush the hydrogen cavities to concentrate energy in an inertially confined plasma.

As well as contributing to fusion research,[12] these companies conduct experiments that produce massive amounts of data. TAE Technology, for example, claimed to have run over 50,000 experiments by 2021. Already by 2014, it had begun a partnership with Google to apply AI/machine learning to accelerate its research. Between them they created the Optometrist Algorithm, which alongside other developments, has significantly increased the speed of research (Baltz et al. 2017). The algorithm provides a human operator with two alternative experimental settings and associated outcomes, and she/he then chooses which experiment produces subjectively better results. General Fusion's experiments, including its computer simulations, can produce as much as 100 gigabytes of data per day; and is collaborating with Microsoft to analyze its fusion energy experimental results using cloud-based big data techniques and machine learning. This, the company says, will enable it to explore experimental plasma data in greater depth than previously possible.

These relatively small fusion firms have demonstrated advances and have attracted funding from a number of the richest people in the world, long-term institutional investors, and increasingly oil and gas companies and venture capitalists. They have built on new generations of super computers and superconductors, aiming to design and test different ways of producing fusion faster and cheaper than large government-funded research projects. They have also attracted partnerships around data science. But some observers express deep scepticism about the claims made by such fusion startups, calling them "unscrupulous"[13] and even comparing them to Theranos, the fraudulent health technology company (which itself attracted significant funding and an eminent Board of Directors).[14] One of the most sceptical observers (Siefe 2008) claims that the one real, concrete success of fusion startups is not extracting energy from seawater, but extracting money from speculators.

There has undoubtedly been hyperbole in the selling of the business model of small fusion firms although the "fake it until you make it" approach is not uncommon in technology startups. The sheer scientific and engineering complexity of producing fusion energy—the need for extensive scientific expertise, the manufacture of complex new components and systems, and the ability to integrate design and construction—far exceeds the capacity of small firms to address it. Producing the extreme heat to produce fusion reactions is but one of the problems to be addressed, and just as challenging is converting the heat into usable power or electricity.

Beyond unmet claims, these firms can alternatively be judged by their contributions to their fusion innovation ecosystem. This may take the form of contributions to fusion technology, as seen, for example, in the case of TAE Health. It can be seen in Tokamak Energy and its developments in magnets. And it can be seen in the ways their various experiments add to knowledge of alternative approaches to generating

[12] See, for example, https://firstlightfusion.com/science-hub/validate-production-of-neutrons-from-gas-gun-driven-targets.

[13] "Former fusion scientist on why we won't have fusion power by 2040" https://www.youtube.com/watch?v=JurplDfPi3U.

[14] See Carreyrou (2023) and The Theranos trial shows why we need to be suspicious of nuclear fusion. (slate.com).

fusion energy, and, importantly, insights into what does not work. Many firms which set out to produce fusion energy may in the end become important contributors of specialized products and services to industry supply chains. There are likely to be more firms emulating First Light Fusion as it changed its strategy—and amended claims that it would itself produce net power by 2024—to become a supplier of expertise and components. As a general contribution their entrepreneurial spirits may also infuse and amend the approach to innovation of much longer established research institutions.

What of large industrial firms? For fusion energy to be viable, it will need to be produced at scale, with potentially hundreds if not thousands of machines in operation. These, in turn, will need to be integrated in an energy system for distribution and storage. The ability to do this resides only in construction and engineering companies with the deepest pockets and greatest expertise. Outside of the defence industry, the pool of such countries in the West is likely to be limited.

One of the most important stimuli to the fusion innovation ecosystem will lie with the demand of leading energy-intensive businesses, including, importantly, data and cloud service centres. The Cloud is estimated to have a larger carbon imprint that the airline industry, and a single data centre can consume the energy needed for 50,000 homes.[15] In a 2025 report, the International Energy Agency estimates that electricity demand from data centres worldwide is set to double by 2030 to around 945 terawatt-hours, slightly more than the entire electricity consumption of Japan.[16] This has led companies, such as Google and Amazon, to enter into agreements to produce small nuclear fission reactors alongside data centres.[17] With the further development of fusion machines, their advantages over fission machines may become more apparent to such large and discerning customers.

7.3 UKIFS/STEP as an Innovation Ecosystem Catalyst

Efforts to intensify fusion innovation ecosystems are occurring around the world, and the UK's STEP programme provides an example. STEP is an initiative with the intention of hastening and increasing interactions between research and industry; essentially animating and intensifying the fusion innovation ecosystem.

Discussions about creating STEP began amongst the leadership of the UKAEA around 2018, and were conducted in the context of continuing difficult negotiations around the UK's funding of EU fusion research, the increasing number of private

[15] hereader.mitpress.mit.edu/the-staggering-ecological-impacts-of-computation-and-the-cloud/#:~:text=The%20Cloud%20now%20has%20a,as%20much%20as%2025%20percent.

[16] https://www.iea.org/news/ai-is-set-to-drive-surging-electricity-demand-from-data-centres-while-offering-the-potential-to-transform-how-the-energy-sector-works.

[17] https://blog.google/outreach-initiatives/sustainability/google-kairos-power-nuclear-energy-agreement/. https://www.aboutamazon.com/news/sustainability/amazon-nuclear-small-modular-reactor-net-carbon-zero.

sector startups in the USA, and the success of MAST-U, demonstrating the feasibility of managing the control of heat in smaller fusion machines. It was felt there was momentum in fusion energy, and the question was how the UK, and the UKAEA, were going to respond.

On his first visit outside of London following his election as Prime Minister, Boris Johnson toured Culham on ninth August 2019, and celebrated its world-leading research.[18] For decades UK governments of all political persuasions, and facing all circumstances, had consistently supported science and its potential to create new industries and jobs, and Johnson was interested in how to exploit fusion research. He was presented with a proposal to design a new spherical tokamak machine in Britain as a prototype to make power, learning from JET and MAST-U. The decision was made in 2023 to allocate £42 million to the proposal in a Fusion Industry Strategy, part of an overall funding package at the time of £405 million.

Initially a wholly-owned subsidiary was established within UKAEA, with plans to latterly create an organization to build and own STEP and to receive external investment. A great amount of attention over 18 months was devoted to planning what the best institutional model was to be, with advice taken from a group named UK Government Investments and the Treasury, organizations with extensive experience of privatization. Examples were sought from other sectors, such as defence. On the seventh February 2023, Minister of Science George Freeman announced the delivery body for STEP was named UK Industrial Fusion Solutions (UKIFS).

On the 31st March 2023, then Prime Minister, Rishi Sunak visited Culham with Grant Shapps, the Minister of Energy. There was still uncertainty at the time about the UK's position on European fusion programmes for which funding had been appropriated. During this visit, the case was made that these funds would be allocated to more effect to a British fusion prototype. As a result, the government committed £220 million to fund the first phase of STEP; and see the UKAEA produce a concept design for it by 2024. The aim is to have a fully evolved design and approval to build by 2032, so that construction can begin. UKIFS was charged with meeting four transition measures to enable it to operate in full as a subsidiary and develop a Target Operating Model, which UKIFS would utilize to develop and build STEP in full capacity. UKIFS reports to the Department of Energy Security & Net Zero, and as the developer of a major project has also to account to the government's Infrastructure and Projects Authority, tasked to ensure the efficiency and effectiveness of major government projects.

UKIFS's core activity lies with being an integrated delivery team, working with three specialisms: technology, construction, and engineering. As the construction and engineering partners will need to bear significant financial and technical risks and require deep and wide expertise, they are approached as consortia. As of 2025,

[18] https://www.gov.uk/government/news/prime-minister-hails-world-leading-fusion-research-at-culham.

the exact forms of providing the construction and engineering specialisms, and the involved partnering organizations, are still being determined.

To proceed through its approval processes, the UKIFS had to demonstrate its basic capacity to function, notably to employ people, and care for their health and safety. Its governance is established with a Board and its audit, risk assurance, and "people and remuneration" committees. It has an international scientific advisory team of 15 leading scientists.

These decisions to support STEP were made during a particularly tumultuous period in British politics, and where the authority to make them was occasionally contentious. Decisions were also being made in a public service not renowned for its expertise in science, technology, and industry[19] although the quality of the public servants working on this project was and remains widely applauded. The fact that the project proceeded is also testimony to the strength of the case, and its effective lobbying, and the commitment of key politicians, including three Prime Ministers and a number of Ministers.

A significant challenge lay in creating regulations fit for fusion energy production, rather than fission power stations, where most regulations were directed. A successful case was made that as a safer technology fusion did not need to be legislated under nuclear fission regulation and licensing requirements. After a public consultation and the production of a government Green Paper in 2021, legislation to create an independent fusion regulator passed through Parliament in 2023. Since that time, fusion energy in the UK is independently regulated by the Environment Agency (EA) and the Health and Safety Executive (HSE). One immediate benefit of this decision was that STEP did not need to be built on a previously approved nuclear fission site, and it could be located on a site not associated in the public's mind with nuclear fission. It also avoided the nuclear fission regulations that burden ITER.

A competition was created to invite sites for locating STEP, and 15 proposals were received from throughout the UK. A key consideration was the need for a cooling water license, so the site had to be on the sea or on a significant river. Another requirement was access to the phenomenal amount of power needed to start the fusion reaction, and to a large and reachable community of scientists and engineers. A large site was needed to build all the ancillary industry required in suppliers and for maintenance and to meet the ambition of creating training centres and innovation laboratories around the plant. A cross-functional team at the UKAEA, supported by multiple sources of relevant expertise, undertook due diligence and

[19] Professor Sir Patrick Vallance, who was Chief Scientific Advisor to the government before becoming Minister for Science, has called for more scientific expertise in government (https://www.theguardian.com/politics/2020/jan/12/whitehall-needs-more-scientists-to-compete-with-china-patrick-vallance-dominic-cummings), a conclusion echoed in a 2019 government review of science capability in government. One challenge lay in demands to assess detailed designs for STEP before approval to proceed with appointing the engineering partner which would be central to the design. A refrain heard in government was that the risk of STEP was too high as there were no equivalent projects to calibrate it alongside. As a first-in-the-world project, this is unsurprising.

engagement with local communities at all the sites. Over a period of 3 years, they were narrowed down to five and then to two. In 2022, George Freeman decided STEP should be located at a large industrial site in West Burton in Nottinghamshire. The 300-hectare site had the benefit of being on the navigable River Trent with a water license, and as an ex-coal-fired power station it had access to the power grid and a railhead for transportation.

UKIFS has two missions:

> to build the world's first demonstration prototype fusion plant and to inculcate and foster a fusion industry

In short, to build a machine and build an innovation ecosystem. In practice, UKIFS aims to be an intelligent client at the centre of the intellectual property around STEP, such that this creates a supply chain around the design, engineering, and construction of the fusion machine and the plant required to operate it.

One concern behind creating STEP was to speed up the development of fusion energy by encouraging researchers in the UKAEA to think beyond their scientific experiments in areas such as refining magnets, improving materials, managing unstable plasmas, and advancing sensors and digital models, and building a commercial machine as a focusing mechanism to bring this expertise together. There was a view amongst some researchers in UKAEA that STEP was premature—there were still scientific questions to be resolved—and considerable adjustment is needed in shifting from the design and maintenance of a scientific machine for research to a commercial machine.

Paul Methven summarizes the approach taken by STEP and its role:

> The major underlying challenges of fusion have been approached from a scientific perspective, which is we need to find a way of understanding and modelling the plasma, we need to find a way of understanding the heating operation, how to breed tritium, we need to understand a way of providing heat into the system. We need to understand an approach to maintainability and so on and so forth. And it's perfectly rational to say, at some level, those are pretty fundamental challenges and they need to be approached bottom up. What we are saying is, fine, we get that, but let's do this the other way around. Let's say the end point of this is an integrated plant design. And by trying to do that, even when those subordinate elements are not resolved, what you actually do is give a better definition to the problem statements at that lower level. So we're applying some much clearer constraints and parameters and definition to that problem by saying it's not just any old solution to tritium or any old solution to materials. It has to be this, it has to balance these factors. It has to fit this set of temperature, pressure limits. It has to resolve this interface problem. Our approach is deliberately structured to produce an integrated plant design and that forces people to work together in interdisciplinary groups, to resolve those problems

The design of a commercial tokamak is very different from a research machine. The energy created in the vessel, for example, will need to run for longer periods. As experiments only run for matters of seconds, the operating costs of commercial machines will be much higher. As research machines, tokamaks have numerous sensors probing into the vessel, which will be unnecessary in a commercial machine. The "real estate" around the machine will differ, with less emphasis on research instruments and more on heat capture. In large tokamaks, such as the one being built at ITER, the plasma is further away from the walls than the one planned at STEP, where its small, spherical shape makes exhausting the heat more challenging. MAST-U demonstrated that it was possible to make such a tokamak without melting it using less power and fewer magnets, but one has yet to be built. The challenge remains to create a blanket that captures the enormous energy from the fusion reaction and which also produces the tritium that provides the machine's fuel.

There are opportunities for innovation including exploring new methods for heating the vessel, which in JET used a system known as neutral beam injector that injects high-energy neutral particles. This takes up a lot of the machine's "real estate", which could otherwise be used for heat capture, and the UKAEA is currently researching the ability of microwaves which could in principle be directed from hundreds of metres away. The group researching microwaves have applied for numbers of patents and there are opportunities for industrial applications outside of fusion.

Lessons from JET are being used in the design of the building in which STEP will be located. When JET was operating, it was encased in a chamber with two of the largest and heaviest doors ever constructed. This was necessary for safety, but posed massive challenges for maintenance where large components needed to be lifted in a gantry crane and moved out into an antechamber to be worked on. Rapid maintenance will be crucial for the commercial operation of a fusion machine, whose economics depend on its continual running. The heat divertors on the STEP machine make its total size likely to be around 25 m high, which if access for a crane through safety doors is necessary, makes for a very large and expensive building. This challenge also provides opportunities for innovation in design and construction.

One area for innovation is modularization of the components of STEP, so that smaller parts can be lifted out on a gantry crane. This involves, for example, demountable magnet joints. The production of high-temperature superconducting magnets is highly complex and the challenge of designing them so that they can be separable is a multi-technology design challenge which is currently being researched in the UKAEA and the fusion companies, Commonwealth Fusion and Tokamak Energy, and the magnet company, Oxford Instruments. The UKAEA has developed a new technology to join magnets together that is also more environmentally friendly than existing techniques. Separable, modular, "unpluggable", high temperature superconducting magnets would have significant demand outside of fusion in other applications such as energy distribution, generators, and high field magnetics.

Simply, the approach being taken is that the lid of the vessel could open to allow access to a crane that would remove the central core and move it to a work hall for maintenance and replace it with a spare core to ensure maximum operating time. This and the maintenance would be robotized, so no human is exposed to radiation or heat.

This and the whole operation of STEP is developing a fully immersive digital twin, so much on the necessary training and testing can be undertaken *in silica*. STEP has benefitted through the ability to simulate and model using very sophisticated digital models. These provide greater confidence in foreseeing and overcoming technical obstacles. A number of design principles are embedded, including design for manufacture, ensuring that the ability to economically construct the prototype is factored in alongside design options; planning for several generations of prototypes, each building on the learning from predecessors; and preparing for related benefits to accrue if designs fail. The whole project is benefitting from technological developments in the fields of advanced manufacturing and 3D printing, and in the use of AI.

Paul Methven summarizes some of the philosophy behind the design of STEP:

> We've made lots of progress on the underpinning technologies by defining challenges more clearly and then starting to do progressive developments and demonstrations on them. So I think we've got a much better handle on what we need to demonstrate over the next few years. And that's starting to create not just technical solutions, but an increasing portfolio of innovation and intellectual property. And as we get engaged with our supply chain more and more, then that will breed innovations from them too, not just in fusion, but in wider fields.
>
> It's absolutely implicit in the programme that if somebody else makes a breakthrough or we find that we've got a dead end, and we need to pivot to some other basis, that we take the capability and the learning that we've developed and we apply that to something slightly different. We have to be humble enough to change, and we have to be agile enough to do so.

These aims complement the industry strategy of the local region, which has a task force comprising the UKAEA, UKIFS, and the Midlands Engine.[20] The latter is charged with attracting greater investment into the region, where there is already a high value manufacturing cluster that includes companies such as Rolls Royce, Boeing, and McLaren. There is also a regional plan for energy transition following the decommissioning of 13 coal-fired power stations along the River Trent Valley, involving efforts to attract energy transition R&D companies. The decommissioning of the power station West Burton was part of the UK's ending in 2024 of 140 years of power produced from coal, and signals its search for alternative sources of energy. Fusion fits this strategy exactly and it benefits from the industrial supply chains in existence being fostered in the region.

An example of the sort of development associated with STEP is provided by CHIMERA (Combined Heating and Magnetic Research Apparatus), a fusion component testing facility in Rotherham, South Yorkshire.[21] Located at the UKAEA's Advanced Manufacturing Park, 30 miles away from West Burton, it tests and develops materials and components and simulates hydraulic, thermal, electromagnetic, and mechanical systems in the extreme conditions of a fusion power plant without nuclear reactions. Much still needs to be learnt about how materials and components will behave in a fusion machine when it is operating beyond the short pulses of a research machine. This includes how composite materials and welding joints deal with extreme stresses.

CHIMERA is aiming to develop very distinctive capabilities that help answer these questions. It is the only device in the world to test fusion component prototypes in a vacuum environment with high heat and magnetic fields; and furthermore

[20] https://midlandsengine.org/.

[21] https://www.rothbiz.co.uk/2021/04/news-7685-chimera-makes-rotherham.html.

can test large component prototypes up to 1.8 m tall. A £22 million investment, CHIMERA has Europe's most powerful civil laser beam to create the energy flux found in a fusion machine. It will be crucial for testing the wall of STEP as it is being designed, and its unique capabilities have attracted a great deal of attention, including from the nuclear fission industry, and there are plans to expand its facilities and testing abilities.

STEP's likely local impact was described in an economic assessment of the programme, undertaken by AMION Consulting in 2025, which estimated that more than 6500 jobs a year could benefit Nottinghamshire residents at the peak of STEP's operations, and that Nottinghamshire's economy could be boosted by between £86 million and £210 million each year, for the following 40 years.[22] It estimated that around 70 percent of the workforce will require high qualifications, ranging from higher apprenticeships to Master or PhD degrees, and that STEP will train 1200 people in skilled areas such as welding, electrical, and mechanical work, and offer 330 apprenticeships during the construction phase of the project.

A journey of exploration, such as that of the pursuit of viable fusion energy, is always capable of producing the remarkable and unexpected. A journey that arguably began in the nineteenth century in Marie Curie's modest laboratory in the Latin Quarter of Paris, that was energized by the building in the 1960s of a tokamak in the Kurchatov Institute in the unassuming Shchukino District of Moscow, that has been nurtured and intensified for six decades in a small, rural village in Oxfordshire, may arrive at some form of destination in the redundant wasteland of an old coal-fired power station in the East Midlands of England.

References

E. Baltz, E. Trask, M. Binderbauer, et al., Achievement of sustained net plasma heating in a fusion experiment with the optometrist algorithm. Sci. Rep. **7**, 6425 (2017)

J. Carreyrou, *Bad Blood: Secrets and Lies in a Silicon Valley Start-Up* (Picador, London, 2023)

I. Chapman, S. Cowley, H. Wilson, The spherical tokamak for energy production: Theme issue introduction. Phil. Trans. R. Soc. A **382**, 20230416 (2024)

T. Clyne, 5 Big ideas for making fusion power a reality. IEEE Spectr. **57**(2), 30–37 (2020)

M. Coopersmith, Fusion is not a typical bet. Bull. At. Sci. **80**(6), 358–364 (2024)

R. Pearson, A. Costley, R. Phaal, W. Nuttall, Technology Roadmapping for mission-led agile hardware development: A case study of a commercial energy start-up. Technol. Forecast. Soc. Change **158**, 1–23 (2020)

C. Siefe, *Sun in a Bottle: The Strange History of Fusion and the Science of Wishful Thinking* (Penguin, New York, 2008)

R. Zhong, *The Quest of Build a Star on Earth* (New York Times, 2024)

[22] Chrome extension://efaidnbmnnnibpcajpcglclefindmkaj/https://www.nottinghamshire.gov.uk/media/vvhf5poy/stepsummaryreport.pdf.

Part IV
Paths

Chapter 8
A Story Still in the Telling

The story of fusion energy continues: stories don't always have endings, and certainly don't always have happy endings. But stories are often about journeys, and our purpose has been to tell of the pursuit of viable fusion energy. It has involved a journey of people, including theoretical scientists concerned with fundamental physics, technologists, and engineers building incredibly complicated machines, and policymakers and businesspeople aiming to create new systems for energy production and distribution and ways of delivering energy for public good and private profit. It is an expedition of learning and knowledge accumulation, a story of the journey of fusion energy's innovation ecosystems as it coalesces and builds the effective connections between all the resources and skills needed to make fusion energy viable.

8.1 A Journey Worth Making?

The question most often asked of fusion energy is when will it become a reality: when will the story end? The answer is that it is already a reality and the quest will never end. The incredible achievements to date, of replicating stars of superheated plasmas, is testament to the extraordinary capacity of humankind to direct vision, curiosity, and resources to solve problems. The achievements have yet to reach the target of delivering energy that is affordable, accessible, and sustainable, but we are approaching the situation where viability is attempted to be demonstrated in pilot plants. There is, however, still be a long way to go if these are successful to achieve the scale necessary in the number of fusion energy plants to positively impact climate change. It is unlikely that fusion will make any significant contribution to global energy supply before 2050. The UKAEA estimates that after 2050, fusion

M. Dodgson, D. Gann, *The Search for Viable Fusion Energy*, SpringerBriefs in Applied Sciences and Technology, https://doi.org/10.1007/978-3-032-13466-0_8

could reach 10 percent of global energy demand by 2100, or as much as 40 percent if capital costs were 30 percent cheaper.[1]

The next stage in the journey will require addressing the huge challenge of producing commercially viable machines and building new supportive industries to begin to address the problems of carbon-producing energy sources. New technologies and ways of making and organizing will emerge, as they have in the past, changing the path of the development of fusion energy in ways that are presently unseen. The capacities of AI, for example, to assist in the design of fusion machines and optimization of energy distribution and control systems are high potential, but still emerging. They are crucial for the development of prototype machines such as STEP and China's CFETR. Assessing whether the journey has been worth it needs also to assess the development of fusion-related technologies as well as fusion energy. Serendipity has a way of delivering unexpected benefits, and fusion technologies may conceivably produce far greater returns than the costs of developing fusion energy.

Significant challenges remain to be faced if fusion energy is to be viable. In tokamaks, for example, these include the production of the feeder fuel, tritium, the materials used on the walls of vessels, heat management, and the ability to manage the quench phenomenon of high-temperature superconducting magnets. Apart from the complexity of the science and engineering problems still to be solved, the question remains about the economic viability of fusion energy. How much will fusion plants cost? What amounts of energy will they generate and at what price? These questions have to be posed alongside addressing the objections of those opposed to all forms of nuclear energy, proponents of alternatives such as solar and wind who argue these are all that are necessary, and opponents concerned about radioactive waste and military applications.

All these questions have to be asked in the light of whether the world can afford not to invest in fusion energy, given the rapid rate of climate change and failure to develop sufficient scale of viable alternative sources. Any answers to the questions of whether the pursuit of fusion energy is worth it have to account for the way that the challenges of delivering carbon-free energy are astronomical, and the cost of not pursuing them is infinitely higher. Every kind of alternative, many forms of renewables, nuclear, carbon capture and storage, and production of clean hydrogen are likely to be needed to meet net zero aims and obligations. The depressing fact remains, however, that around 0.4 percent of global GDP is being spent on climate change, whereas the amount required is at least 4.0 percent.[2] Government initiatives, such as ITER, may appear expensive, but its cost over a number of decades is equivalent to the annual profit of most major oil companies. Total private sector investment in fusion firms to date is about 0.3 percent of the amount of venture capital invested in 2024. This lack of attention may reflect the view that fusion

[1] chrome-extension://efaidnbmnnnibpcajpcglclefindmkaj/https://committees.parliament.uk/writtenevidence/137642/pdf/.

[2] Interview with Ian Chapman, CEO UKAEA, 17/12/21.

remains seen as high risk and long term, or that the challenges to be faced have yet to be properly appreciated or calibrated, or indeed, that fusion is not relevant for this type of funding and needs alternative means of financing. One estimate calculates $3 trillion is expected to be invested in building AI data centres, with Google, Amazon, Microsoft, and Meta spending $750 billion in 2025/26.[3] Another estimate predicts data centres will require $7 trillion of investment by 2030.[4] If these resources can be found for highly energy-consuming projects, the question for public policy, strategic decision-makers in business, and society more broadly, is to calibrate the level of such investments compared to those that offer the potential to produce new sources of environmentally sustainable energy.

It does need to be recognized that the timescales involved when ideas emerge from basic research are lengthy. Fusion research is often castigated for the length of time it has taken to deliver practical results, but such timelines are not uncommon: the electric car was developed and demonstrated in the mid-nineteenth century, and it took over 150 years for it to become widely diffused. The electric vehicle demonstrates how it is not simply scientific and technological obstacles that slow innovation. Also to be included in the comprehension of "time to market" are the complementary industries and assets that surround an innovation, such as the technical challenges of battery production and recharging, the powerful effect of incumbent interests and infrastructure, and obstructions from those who benefit politically from opposing change.

We now turn to some of the major insights, questions, and challenges of the pursuit of fusion energy from the perspectives of the contributors to fusion innovation ecosystems.

8.2 Research and Research Institutes

Much of human progress has been built on the ability to turn ideas into valuable uses. Especially at its early stages, research is an experimental process around ideas, creating options for the future. The development of fusion energy is an epic experiment, and indeed ITER is currently building a scientific instrument that could become the world's largest experiment. Research machines, such as JET, are used to gain new knowledge, and test what works and what does not. They can involve exploring the unknown. Failure, or the lack of success at meeting predetermined objectives, is an inevitable outcome of research and innovation, and it is a valuable contribution to eventual success. Knowing what doesn't work can be as useful as knowing what does. Project ZETA, discussed in Chap. 5, is often portrayed as a failure; yet, it was invaluable in developing fruitful knowledge for subsequent

[3] https://www.ft.com/content/efe1e350-62c6-4aa0-a833-f6da01265473.

[4] https://www.mckinsey.com/industries/technology-media-and-telecommunications/our-insights/the-cost-of-compute-a-7-trillion-dollar-race-to-scale-data-centers.

experiments. Assessing where we are now must account for successes in knowledge creation, even if they are publicly perceived as failures.

Public research institutions such as UKAEA and PPPL have played the "anchor" roles in fusion innovation ecosystems to date, and the question is how ecosystems embed their and other's research as they develop. This includes connecting upstream and downstream innovation with industry. It includes the commercialization of their research and the creation and support of spin-off companies. It also involves the transfer and development of skills in industry and the involvement of what sociology, and science and technology studies, refers to as "boundary objects" and "boundary spanners". Boundary objects include arrangements such as demonstration plants or procurement development workshops, designed to build communication and understanding across multiple "fields", such as sectors, professions, and disciplines. Boundary spanners are leaders and managers skilled at working across fields, developing collaboration and shared knowledge. Such people are essential as organizational cultures change in line with altered purpose. So as the transition to industrial applications proceed, scientific organizations whose cultures are devoted to discovery and experiment have to adjust to engaging with organizations concerned with commercial deadlines and financial outcomes. People whose experience lies with designing machines for experimental science have to adjust to the work of designing and building a working whole-plant prototype. In this way STEP and other prototypes being or to be developed are large-scale boundary objects needing boundary spanners of significant scale and range.

The study of research and innovation uses the concept of dominant designs and explores the process by which they emerge. One bet for a dominant design in fusion energy to emerge from research is the tokamak, but competing and related designs are being tried. The value of a dominant design is the way it focuses investment, concentrating the creation and diffusion of valuable information. The tokamak has provided such a focal device, and variants such as spherical tokamaks are being developed. The existence of a dominant design does not preclude alternative explorations, and fusion research valuably encompasses inertial devices and stellarators and a variety of different approaches in startup companies. On the current point of scientific maturity, a key role for research and research institutes is the development and testing of fusion machines that emerge as dominant designs to move the industry up the S-curve.

8.3 Government

The story of the pursuit of fusion energy has to-date been mainly one of government investment. Although private sector firms are entering the field, their investments are significantly smaller than that of governments. Pertinent questions include whether government investments have paid off, what have governments learned from their investments, and can any lessons be transferred to other emerging technologies and sectors?

Measuring the performance of private sector investments is relatively straightforward and will rely on the typical corporate indicators, including survival rates, technological achievements, and financial performance, such as the ability in their early stages to attract investment and eventually make a profit. Assessing the returns to governments, however, can be more complicated and they always face pressure to justify their public expenditures. Government concerns lie with scientific and social, as well as economic, performance. More generally, policymakers need to consider how fusion research fits into national research and innovation systems, and how it provides valuable options for the future of society. The value of the option of fusion energy in the face of a catastrophic climate crisis would be seen in the terrible costs of its absence. Insights can be derived from the more general study of the returns to basic research and "Big Science", but research in these areas highlights how complicated evaluating performance can be in this kind of endeavour. It is pertinent, however, to examine the returns to government investments in fusion energy to-date as this will influence future decisions.

The lessons from investments in basic research in general, and from studies of fusion research centres, is that a multiplicity of indicators of performance are needed. Scientific performance can be assessed if sufficiently nuanced to allow the broader benefits to the scientific community and industry to be captured. Economic and industrial performance can be evaluated, provided that the "harder" indicators of industrial engagement, such as procurement contracts, licensing, and the wider applications of knowledge, are complemented by the "softer" indicators such as improved learning and social networking capabilities.

Major institutions such as UKAEA and PPPL conduct experiments concerned with advancing knowledge, characteristic of basic research. The benefits of basic research are wide-ranging. Salter and Martin (2001) argue its economic benefits are both real and substantial and take a number of forms, including: 1. increasing the stock of useful knowledge; 2. training skilled graduates; 3. creating new scientific instrumentation and methodologies; 4. forming networks and stimulating social interaction; 5. increasing technological opportunities and diversity and the capacity for scientific and technological problem-solving; 6. creating new firms. As a result of this diversity, they contend no simple model of the nature of the economic benefits from basic research is possible.

Salter and Martin also refer to some of the behavioural and managerial benefits of basic research. These include the way skills are developed in networks to improve the capacity of organizations and countries to absorb externally sourced knowledge, enhance the ability to translate knowledge into practice, and how these are assisted by the mobility of researchers from scientific institutions into industry. Our story of JET and the UKAEA, and the research conducted, training and skills developed, instrumentation created, networks formed, new technological options opened, and spin-offs assisted, show that fusion research essentially delivers all the economic benefits of basic science identified by Salter and Martin. Any assessment of the returns to the public of government investment in fusion research has to account for these factors.

Having referred to the difficulties of measuring returns to basic research, a report by London Economics published in 2020 on the impact of UK investment in fusion, cited in the UK government's *Towards Fusion Energy 2023* strategy, identified a total economic impact to the UK economy of around £1.4 billion in relation to £350 million of public investment in UKAEA's fusion research over 10 years; and that this had supported around 36,000 job years over the same period. In evidence presented to the UK Parliament, the UKAEA estimated the gross value-added impact of UKAEA fusion R&D activities between 2009/10 and 2023/24 was £1.8bn-£1.9bn, with UK government spend of £893 m over the same period—representing a Benefit-Cost Ratio of 2.[5] It also reported the UK economy has so far gained more than £650 million from contracts with ITER.

One other observation from Salter and Martin makes the point that in any collaboration it is important to bring something to the party. The best international scientists will wish to work with their peers and will not tolerate free-riders. As they put it, the capability to understand the knowledge produced by others can only be developed through performing research.

Thus the absorptive capacity of organizations, so crucial to benefitting from external knowledge, relies on internal investments in knowledge. Investments are made not only to search for particular results, but also to improve the ability to appreciate and integrate knowledge from external sources. With the international perspectives of the development of fusion energy, it is those nations that invest in fusion research themselves that gain most from international research efforts.

From the industry perspective it is those firms that invest in fusion research themselves that are best placed to benefit from the research undertaken by others. Firms that invest in their own R&D develop the important "absorptive capacity", to benefit from research undertaken by others (see, for example, Zahra and George 2002).

Consistency in government support matters, and we have seen the historically regular institutional realignment of UKAEA and inconsistent funding for PPPL. We have also seen the benefits of long experience in, for example, the UK's accumulated knowledge of nuclear decommissioning, and the immense contribution of JET during its 40-year operation. While there is continued support for fusion energy in Europe, Japan, and China, the question of the USA's commitment remains unanswered following a new Presidency in 2025[6].

[5] chrome-extension://efaidnbmnnnibpcajpcglclefindmkaj/https://committees.parliament.uk/writtenevidence/137642/pdf/.

[6] https://www.gov.uk/government/publications/uk-research-and-development-roa.

8.3.1 International Perspectives

The enormous scale, breadth, and depth of science often requires international collaboration. CERN, the International Space Station, Square Kilometre Array radio telescope, and ITER, require a level of resources, scientific expertise, equipment manufacturing capability, and data processing capacity that no single country can possess, let alone afford; even the largest nations, the USA and China, or collaborative Europe. International collaboration not only offers scale in the resources for science, but adds to the diversity of its inputs. The different scientific cultures and capabilities across nations produce the diversity of thought necessary to stimulate imaginative new departures in science. Multinational teams, offering depth of expertise across a wide range of disciplines, are needed to deal with complex systems effects, where the integration of various elements of systems have emergent and unpredictable properties.

Similarly, the collection, analysis, and interpretation of data can vary nationally, and the data itself, whether on climate, environmental damage, energy uses, or disease and genetics, varies geographically and improves opportunities to develop robust answers to scientific problems. The growing capacity of eScience, the digital infrastructure for research, facilitates collaboration across borders, as seen, for example, in the highly internationalized analysis of data produced by CERN. Furthermore, the size and complexity of fusion machines require production capabilities in globally integrated industrial supply chains. Components of ITER's tokamak are built around the world, and STEP will unavoidably require internationally sourcing.

The large, multi-disciplinary scientific and industrial efforts needed to deal with the existential threats confronting the world exceed the capacities of individual nations to address them. But despite the obvious benefits of international scientific collaboration, we have to be cautious of increasing techno-nationalism. In a number of countries, international agencies in many fields—defence, trade, science—are facing the challenges of increasingly introspective governments who can be influenced by the way science itself is questioned by sceptical and powerful voices.

In the face of increasing nationalism and protectionism in some leading countries, the innovation ecosystems surrounding fusion energy are highly international. JET was a European project, and ITER involves countries which are experiencing great tensions in other areas. Although fusion energy is still far from being applied in electricity markets, limiting commercial conflicts over intellectual property rights, ITER operates under the principle of equal access to IPRs, which reveals a core belief in internationalism. Scientists from all around the world enjoyed access to JET and will eventually if all goes to plan to ITER—and fusion energy largely exemplifies this necessity. At a time where international tensions are so high, every means of building cooperative relationships are valuable, and science in general and fusion in this instance add to this benefit.

8.4 Innovation Ecosystem

Fusion energy represents a classic case of emerging innovation ecosystems involving scientists in research laboratories and universities, governments as investors and regulators, and large and small firms. Due to the scale of the scientific and technological challenges, however, it is large, publicly funded organizations that have led the development of fusion energy. The next stage of the development of the fusion energy innovation ecosystem will of necessity involve greater action on the part of industry. The task, essentially, if fusion is to be viable, is to create a fusion industry.

The role of the entrepreneurial startup features centrally in Schumpeterian analyses of innovation. Certainly, startups in fusion research have provided diversity in experiments and paths to market and a stimulus to the field. The innovation literature refers to the value of combining the advantages of large firms, primarily their resources, with those of small firms, primarily their behaviours and specialisms, such as their ability to entrepreneurially respond quickly to new opportunities. The industry components of the fusion innovation ecosystems tend currently to be more research-oriented and risk-tolerant small startups, with large companies mainly contributing to the design and manufacture of equipment and facilities. The fusion innovation ecosystem will become embedded once large companies with significant fusion interests become key elements: that is, they become central players in a fusion industry.

This includes, on the supply side, large-scale construction and engineering firms with systems engineering capabilities, all with the belief in the economic prospect of fusion, and depth of knowledge and patience to bring long-term investments to fruition. On the demand side, it involves large customers, potentially those operating large data and Cloud facilities, intensifying the ecosystem through their demanding requirements, as already seen in their participation in large prototyping projects and investments in small fusion firms. This progress in the innovation ecosystem from research to industry has implications for the approaches taken by governments.

One of the most enduring discussions about investments in science and research is the extent to which they should be directed towards fundamental discoveries and understanding driven simply by curiosity and with no practical applications in mind, or by clearly defined uses and purposes described as applied research. Within science, technology, and innovation policy, there has also been an increasing interest in "missions", sometimes called "grand challenges". The UK government's Research and Development Roadmap (2020), for example, refers to the importance of "moonshots"—coordinated, multi-disciplinary challenges with ambitious, measurable goals that will have a significant impact on an important societal issue. Another feature of policies in this field lies with the assumption that there is linear progression from basic research to industrial innovation: researchers in universities and laboratories discover new knowledge, which industries apply. The development of the fusion energy innovation ecosystem offers insights into these issues, and for the development of future government policies for science, technology, industry, and innovation.

Contemporary fusion research is mission-oriented, but is also discovery-based. Fusion has moved from deep science discovery/curiosity through mission-oriented research, to the development of commercial prototype in a 70-year period. Its mission is to develop fusion energy to demonstrate viable fusion energy without causing environmental damage. In the pursuit of this mission, numbers of discovery-based or fundamental research questions are being addressed, such as understanding the behaviours of plasmas and materials in extreme temperatures. Fusion research belongs to that type of research described by Donald Stokes (1997) as Pasteur's Quadrant, with clear objectives in mind, but also concerned with explorations aimed at fundamental understanding. It provides insight into how this balance can evolve over time.

In that sense, a fusion energy innovation ecosystem reveals the time dynamics in Donald Stokes' quadrants. Although it includes elements of basic and applied research, discovery and mission policies, over time in both cases their centres of gravity move. As we've seen, however, interactions in the fusion energy innovation ecosystem is not simply single directional. Research does not produce pieces of results that are once and for all handed over to industry: there are continuous interactions in all directions. While the linear science to innovation model has been repeatedly discounted, its simplicity continues to appeal, and like a vampire rising from the grave, refuses to die. Reality, as shown in the fusion case, is much more complex. Industry also undertakes basic research, universities also conduct applied research, and the connections between basic and applied research are iterative and multidirectional. This has implications for government policies supporting industry.

The idea of industrial strategy in Western nations has waxed and waned in its political favour. Once derided as fruitless attempts to "pick winners", more contemporary approaches think of it as "placing bets", that is developing options. Industry strategy influences the development of the fusion innovation ecosystem, both at national level, seen in the UK's modern industrial strategy and fusion strategy, and local level, as seen in the case of the region where STEP is to be located.

This points to an important lesson for governments, especially when policies that focus on discovery, often the remit of science ministries, and those of missions, often the remit of industry ministries, are not well connected. The opportunities for misalignment is increased, of course, when other ministries are involved, which is often the case in long-term major projects. The idea of Pasteur's Quadrant is influential in policy circles, and examples such as fusion energy point to the value in replacing this conceptualization with the idea of Pasteur's Space, emphasizing the fluidity of connections in the framework, and emphasizing the importance of cross-portfolio and whole-of-government policymaking.

Fusion innovation ecosystems will develop with different characteristics in different nations. Given the pressing need to produce new energy sources, the question is not whether one configuration is better than another, but to encourage all to perform as efficiently and quickly as possible, and each to maximize their learning from the progress and setbacks in others.

8.5 Cautious and Contingent Optimism

The journey we have told of the pursuit of viable fusion energy has included the stories of institutions, government policies, and business strategies that tell us much about the range and scale of challenges in moving complex ideas from deep science into practice. And it also tells the stories of people and their motivations, remembering Kennedy's dictum that things are worth doing because they are hard: to search and to strive are amongst the best of human attributes.

We have spent more time than others might on human factors, including the unquenchable motor force of curiosity. Of course curiosity is bounded by resources and institutions, but these are finite while the desire to understand is infinite and boundless. A strategy, policy, or mission is tied to time whereas the purpose of curiosity-driven discovery is timeless. What, why, and how things happen are eternal questions humankind is driven to answer, and which are supercharged when applied to existential challenges, such as the one confronting the global climate. The pursuit of the Sun gods is driven not only by physical, tangible need, but by the unquenchable motive of intellectual curiosity. And this curiosity needs to be nurtured and harnessed in organizations with the resources, structure, leadership, and culture such that fruitful progress takes ideas from science into applications.

This, inevitably, leads us to an optimistic view about viable fusion energy, a view which is supported by the historical example of the rapid and successful development of electric power systems during World War I, a time of acute global crisis (Hughes 1983). Hughes shows how political leaders and public pressure led the managers and engineers who were developing electrification in Western societies to attenuate their customary drive for profit and emphasize the "cooperative production of energy". Today's political circumstances are certainly vexed and distressing, but few will argue they are worse than those during the first global war. It has been and can be done; efforts of the scale required have been and can be mobilized.

Our optimism is bounded by a number of contingent factors. This involves broad recognition that given its objective and purpose, and the related benefits that can accrue, the journey taken in the pursuit of fusion energy is not "nice to have", but absolutely necessary. Its realization will depend on whether leaders in politics, business, science, and government will join with public pressure to emphasize the cooperative production of energy and to drown out fatalism with voices attuned to the need to combat the climate crisis.

Another factor is the distinction between scientific and commercial viability. Experimental machines that produce energy from fusion exist and their operational performance will get better. But there is an enormous difference between these devices, limited in scope to the production of scientific data, and the types of machines envisaged to run constantly and reliably to provide the world with almost limitless energy. The greater uncertainty remains that no one has yet produced a fusion device that can deliver sustained net power output, and questions therefore also remain as to whether the costs of fusion energy will be competitive with alternatives. This will depend, of course, on whether the true cost (environmental and social) of using fossil fuels is eventually realized, on the full costs of alternatives, and on the performance of fusion machines. Here, much will depend on continual experimental development of particular machines, learning about and improving their designs in actual power plants, such that dominant designs emerge that can facilitate the economies of scale that determine optimized prices. This process of development has been achieved through iteration with every other major new innovation in power production since the invention of the steam engine; and in each case, it has taken something like 70–80 years before a dominant, "cost effective" source of power has achieved a discernible share of global energy markets.

If fusion energy becomes viable, the bounteous energy of the Sun will be brought to Earth not by mythical gods, but by the reality of people working cooperatively in science and technology, government, and business. The challenges are immense, but, like all the epic journeys in classical mythology, great trials and tribulations need to be overcome before they conclude successfully. Amongst all the factors that will determine whether fusion energy will arrive at a destination that successfully delivers on its potential, we briefly draw on our combined seven decades of research into innovation by highlighting three crucial issues: developments in China; the application of digital technologies; and the quality of leadership and management.

In our book on the development of China's leadership in many areas of innovation (Zhang et al. 2022), we show how China's extraordinary industrial strengths are now being complemented by impressive capacities in science and entrepreneurship. China is the first major economy to develop high tech industries powered by digital capabilities, in research, manufacturing, finance, and logistics. China is also confronted, of course, by the pressing need to develop non-carbon-emitting sources of energy and is capable of sustaining significant investments over long periods of time. In short, China has the ambition and ability to lead in the future provision of energy from fusion and is ramping up building its fusion innovation ecosystem by adding private sector initiatives to its government-led investments.

The question for the rest of the world is to find ways to collaborate with and benefit from the speed and capacities of Chinese innovation and deep and wide expertise in its industrial supply chains. A challenge for many countries is to develop the deep expertise in manufacturing excellence and agility that China possesses. Whether China will be able to develop commercially viable designs of fusion energy machines may also depend on the challenges that competition can bring in areas such as alternative designs and competitive procurement. The path China adopts organizationally and commercially will crucially influence the future of fusion energy.

Our book on the changing innovation process (Dodgson et al. 2005) identifies what we call innovation technology, that is a suite of digital technologies that intensify innovation from the conduct of science to design and engineering. Recent developments in the commercial use of AI, and possibly in future quantum technologies, inject steroids into the technology platforms that support science and innovation. A major question lies in whether, in the modelling of plasmas or behaviour of materials, or the design and planning of fusion facilities, or predicting demands and pricing of energy, the application of AI will profoundly affect and potentially accelerate the progress of fusion energy. Just as AI companies such as OpenAI and DeepSeek emerged so rapidly with such consequence, who is to say a fusion AI company might appear to similar effect. It is certainly the case that the future of fusion energy will rely on extraordinary computer power, such as that of El Capitan in the LLNL, for modelling and predicting designs and their outcomes.

In our books on the management of innovation (Dodgson et al. 2008; Dodgson et al. 2014), we analyse the broad range of strategic leadership and operational management skills needed for successful innovation. These are all needed for the development of fusion energy, but leadership and management of fusion also requires significant "boundary spanning" skills. On the one hand, this requires the ability to work with governments (whose departments often don't speak with one voice), as investors and regulators. It involves painting a persuasive picture that in times of straightened financial circumstances, commitment is needed presently for uncertain outcomes in decades to come. On the other hand, the same organizations need to lead on the coordination and channelling of the expertise of researchers, imbued with the motive of discovery, and technologists set on commercializing their designs, along with the demands of investors in the public and private sectors, requiring profit.

Leaders and managers capable of balancing and iterating between scientific and commercial objectives, managing effectively within the organization and outside of it with stakeholders, will be essential for the future development of fusion energy. A key observation from our study of different approaches to the development of fusion is that leadership, organizational structure, and culture matter in creating the setting to resolve huge technological challenges and risks. There is no one right model for leading the complex system of capabilities, laboratories, industrial suppliers, regulators, funders, and wider communities of stakeholders. Ultimately, this also includes the capability to deliver a major mega-project requiring investment of billions of dollars, involving thousands of companies and tens of thousands of people. Using the lens of innovation ecosystems, the challenge is often the paradoxical need to have institutional stability and organizational transformation: to have security in funding and clarity of objectives, while maintaining the capacity to evolve and pivot.

The UK STEP Programme was deliberately established with an organizational framework that drew on years of experience of major projects, substantive reviews, and academic literature on organizing major projects and programmes to deliver stability and flexibility. This will be one model amongst others that are, and no doubt will be, developed, but all fusion projects will need to be organized and led in structured yet adaptable ways which enable progress by learning and responding to emerging opportunities and challenges. Opportunities exist to learn from the different approaches to developing fusion, be they private sector-led or state-led, or some form of hybrid. But we believe success will depend on using the distinctive contributions of all the players in innovation ecosystems: government, research institutions, and firms large and small, and melding them together by exceptional leadership.

Paul Methven summarizes the leadership challenges of developing a fusion prototype:

> We're trying to develop the technology and design, develop the site, develop the organisation, secure the partners in delivery terms that we need, and evolve that together and scale the whole thing up. And you have to do all of these things concurrently. It's lovely to be able to say we'll do this first and then we'll do that and so on, but we have no choice but to do them all together and that really is effectively the leadership challenge that comes from it.

These are amongst the many other factors, such as availability of finance, effective regulation, and scientific and industrial capabilities, that will determine whether the pursuit of the Sun gods we've analysed in this book will be successful. Despite all these complexities, it is just as well to be optimistic about the journey of fusion energy as we have dedicated this book to our grandchildren. We have little idea of the lives they will lead, but we hope they have access to the energy necessary to do the things they want to do, that they can afford it, and that it does not destroy the world's environment in the manner of previous generations. Despite the formidable difficulties that need to be overcome, we're cautiously optimistic that fusion energy will contribute to making that hope reality.

Great stories and great journeys are inspiring, and all the difficulties that confront the pursuit of fusion energy—technological, economic, political, managerial—are shaded by fusion's capacity to inspire. The search for viable fusion energy is testament to humanity's endless curiosity and desire to understand complex issues and solve difficult problems. The reasons to pursue fusion energy have to include the ways it affirms these most noble of motives and ambitions.

References

M. Dodgson, D. Gann, A. Salter, in *Think, Play, Do: Technology, Innovation and Organization*, ed. by M. Dodgson, D. Gann, A. Salter, (Oxford University Press, Oxford, 2005)

M. Dodgson, D. Gann, A. Salter, *The Management of Technological Innovation: Strategy and Practice* (Oxford University Press, Oxford, 2008)

M. Dodgson, D. Gann, N. Phillips, *The Oxford Handbook of Innovation Management* (Oxford University Press, Oxford, 2014)

T. Hughes, *Networks of Power: Electrification in Western Society 1880-1930* (Johns Hopkins University press, Baltimore, 1983)

A. Salter, B. Martin, The economic benefits of publicly-funded basic research—A critical review. Res. Policy **30**(3), 509–532 (2001)

D. Stokes, *Pasteur's Quadrant: Basic Science and Technological Innovation* (Brookings Institute, Washington, 1997)

UK Government, *UK Research and Development Roadmap* (HM Government, 2020)

S. Zahra, G. George, Absorptive capacity: A review, reconceptualization, and extension. Acad. Manag. Rev. **27**, 185 (2002). https://doi.org/10.2307/4134351

M. Zhang, M. Dodgson, D. Gann, *Demystifying China's Innovation Machine: Chaotic Order* (Oxford University Press, Oxford, 2022)

Index

M. Dodgson, D. Gann, *The Search for Viable Fusion Energy*, SpringerBriefs in Applied Sciences and Technology, https://doi.org/10.1007/978-3-032-13466-0

Made in the USA
Monee, IL
13 September 2026

62975223R00079